SAFETY SYMBOLS

SAFETY SYMBOLS	HAZARD	EXAMPLES	PRECAUTION	REMEDY
DISPOSAL	Special disposal procedures need to be followed.	certain chemicals, living organisms	Do not dispose of these materials in the sink or trash can.	Dispose of wastes as directed by your teacher.
BIOLOGICAL	Organisms or other biological materials that might be harmful to humans	bacteria, fungi, blood, unpreserved tissues, plant materials	Avoid skin contact with these materials. Wear mask or gloves.	Notify your teacher if you suspect contact with material. Wash hands thoroughly.
EXTREME TEMPERATURE	Objects that can burn skin by being too cold or too hot	boiling liquids, hot plates, dry ice, liquid nitrogen	Use proper protection when handling.	Go to your teacher for first aid.
SHARP OBJECT	Use of tools or glassware that can easily puncture or slice skin	razor blades, pins, scalpels, pointed tools, dissecting probes, broken glass	Practice common-sense behavior and follow guidelines for use of the tool.	Go to your teacher for first aid.
FUME	Possible danger to respiratory tract from fumes	ammonia, acetone, nail polish remover, heated sulfur, moth balls	Make sure there is good ventilation. Never smell fumes directly. Wear a mask.	Leave foul area and notify your teacher immediately.
ELECTRICAL	Possible danger from electrical shock or burn	improper grounding, liquid spills, short circuits, exposed wires	Double-check setup with teacher. Check condition of wires and apparatus.	Do not attempt to fix electrical problems. Notify your teacher immediately.
IRRITANT	Substances that can irritate the skin or mucus membranes of the respiratory tract	pollen, moth balls, steel wool, fiberglass, potassium permanganate	Wear dust mask and gloves. Practice extra care when handling these materials.	Go to your teacher for first aid.
CHEMICAL	Chemicals that can react with and destroy tissue and other materials	bleaches such as hydrogen peroxide; acids such as sulfuric acid, hydrochloric acid; bases such as ammonia, sodium hydroxide	Wear goggles, gloves, and an apron.	Immediately flush the affected area with water and notify your teacher.
TOXIC	Substance may be poisonous if touched, inhaled, or swallowed	mercury, many metal compounds, iodine, poinsettia plant parts	Follow your teacher's instructions.	Always wash hands thoroughly after use. Go to your teacher for first aid.
OPEN FLAME	Open flame may ignite flammable chemicals, loose clothing, or hair	alcohol, kerosene, potassium permanganate, hair, clothing	Tie back hair. Avoid wearing loose clothing. Avoid open flames when using flammable chemicals. Be aware of locations of fire safety equipment.	Notify your teacher immediately. Use fire safety equipment if applicable.

 Eye Safety Proper eye protection should be worn at all times by anyone performing or observing science activities.

 Clothing Protection This symbol appears when substances could stain or burn clothing.

 Animal Safety This symbol appears when safety of animals and students must be ensured.

 Radioactivity This symbol appears when radioactive materials are used.

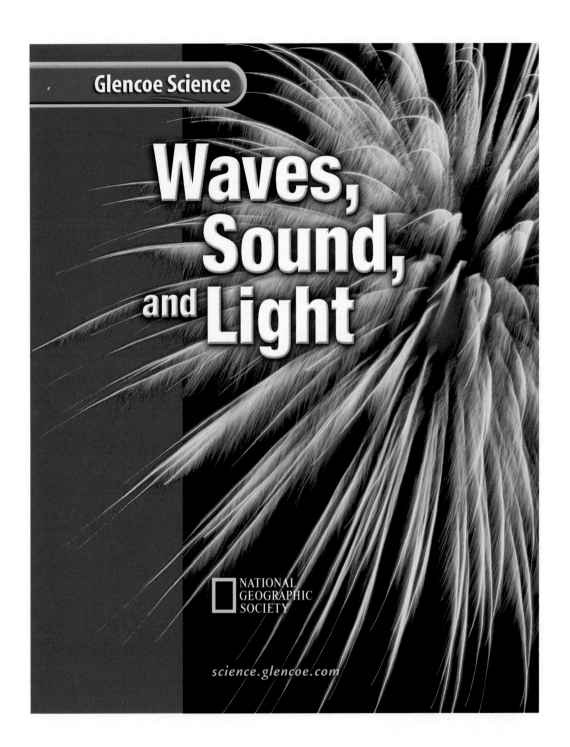

Glencoe Science

Waves, Sound, and Light

NATIONAL GEOGRAPHIC SOCIETY

science.glencoe.com

Glencoe McGraw-Hill

New York, New York   Columbus, Ohio   Woodland Hills, California   Peoria, Illinois

Glencoe Science

Waves, Sound, and Light

Student Edition
Teacher Wraparound Edition
Interactive Teacher Edition CD-ROM
Interactive Lesson Planner CD-ROM
Lesson Plans
Content Outline for Teaching
Dinah Zike's Teaching Science with Foldables
Directed Reading for Content Mastery
Foldables: Reading and Study Skills
Assessment
    Chapter Review
    Chapter Tests
    ExamView Pro Test Bank Software
    Assessment Transparencies
    Performance Assessment in the Science Classroom
    The Princeton Review Standardized Test Practice Booklet
Directed Reading for Content Mastery in Spanish
Spanish Resources
English/Spanish Guided Reading Audio Program
Reinforcement

Enrichment
Activity Worksheets
Section Focus Transparencies
Teaching Transparencies
Laboratory Activities
Science Inquiry Labs
Critical Thinking/Problem Solving
Reading and Writing Skill Activities
Mathematics Skill Activities
Cultural Diversity
Laboratory Management and Safety in the Science Classroom
Mindjogger Videoquizzes and Teacher Guide
Interactive Explorations and Quizzes CD-ROM with
    Presentation Builder
Vocabulary Puzzlemaker Software
Cooperative Learning in the Science Classroom
Environmental Issues in the Science Classroom
Home and Community Involvement
Using the Internet in the Science Classroom

"Study Tip," "Test-Taking Tip," and the "Test Practice" features in this book were written by The Princeton Review, the nation's leader in test preparation. Through its association with McGraw-Hill, The Princeton Review offers the best way to help students excel on standardized assessments.

The Princeton Review is not affiliated with Princeton University or Educational Testing Service.

Glencoe/McGraw-Hill

A Division of The **McGraw·Hill** Companies

Cover Images: A fireworks display lights up the night sky.

Send all inquires to:
Glencoe/McGraw-Hill
8787 Orion Place
Columbus, OH 43240

ISBN 0-07-825630-5
Printed in the United States of America.
1 2 3 4 5 6 7 8 9 10 027/111 06 05 04 03 02 01

Authors

Cathy Ezrailson
Science Department Head
Academy for Science and Health Professions
Conroe, Texas

Nicholas Hainen
Chemistry/Physics Teacher, retired
Worthington City Schools
Worthington, Ohio

Deborah Lillie
Math and Science Writer
Sudbury, Massachusetts

Dinah Zike
Educational Consultant
Dinah-Might Activities, Inc.
San Antonio, Texas

Consultants

Content

Jack Cooper
Adjunct Faculty Math/Science
Navarro College
Corsicana, Texas

Lee Meadows, PhD
UAL, Birmingham Education
Department
Birmingham, Alabama

Carl Zorn, PhD
Staff Scientist
Jefferson Laboratory
Newport News, Virginia

Safety

Malcolm Cheney, PhD
OSHA Chemical Safety Officer
Hall High School
West Hartford, Connecticut

Aileen Duc, PhD
Science II Teacher
Hendrick Middle School
Plano, Texas

Sandra West, PhD
Associate Professor of Biology
Southwest Texas State University
San Marcos, Texas

Reading

Rachel Swaters
Science Teacher
Rolla Middle School
Rolla, Missouri

Nancy Woodson, PhD
Professor of English
Otterbein College
Westerville, Ohio

Math

Michael Hopper, DEng
Manager of Aircraft Certification
Raytheon Company
Greenville, Texas

Reviewers

Desiree Bishop
Baker High School
Mobile, Alabama

Anthony DiSipio
Octorana Middle School
Atglen, Pennsylvania

George Gabb
Great Bridge Middle School
Chesapeake, Virginia

Linda Melcher
Woodmont Middle School
Piedmont, South Carolina

Annette Parrott
Lakeside High School
Atlanta, Georgia

Clabe Webb
Sterling City High School
Sterling City, Texas

Series Activity Testers

José Luis Alvarez, PhD
Math/Science Mentor Teacher
Yseleta ISD
El Paso, Texas

Nerma Coats Henderson
Teacher
Pickerington Jr. High School
Pickerington, Ohio

Mary Helen Mariscal-Cholka
Science Teacher
William D. Slider Middle School
El Paso, Texas

José Alberto Marquez
TEKS for Leaders Trainer
Yseleta ISD
El Paso, Texas

Science Kit and Boreal Laboratories
Tonawanda, New York

CONTENTS

Light, Mirrors, and Lenses—94

Field Guide

Skill Handbooks—134

Reference Handbooks

English Glossary—164

Spanish Glossary—167

Index—171

Interdisciplinary Connections/Activities

NATIONAL GEOGRAPHIC VISUALIZING

TIME SCIENCE AND Society

TIME SCIENCE AND HISTORY

 Accidents in SCIENCE

Science Stats

Full Period Labs

Mini LAB

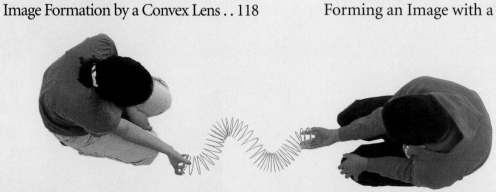

Activities/Science Connections

SCIENCE Online

THE PRINCETON REVIEW

Feature Contents

Let There Be Light

Figure 1
Thomas Edison conducted thousands of experiments to find the proper filament material for one of his greatest inventions—the electric light bulb.

T here's a well-known expression that advises "if at first you don't succeed, try, try again." Fortunately, Thomas Edison lived by these words as he worked in his research laboratories developing over 1,000 patented inventions. Among the many items that his team developed are the phonograph, the first commercial electric light and power system, a motion picture camera, and the incandescent lamp. Edison's search for a suitable filament for the incandescent lamp demonstrates how he used the experimental method to guide his scientific research.

The Search for Filament Material

When electric current is passed through the filament or wire inside the light bulb, the filament heats up and begins to glow. The problem for Edison and his team of researchers was finding a filament substance that would glow for a long time without incinerating (turning to ashes), fusing, or melting.

Before experimenting with filaments, Edison knew that he had to find a way to keep the materials in light bulbs from incinerating. Oxygen is required for a substance to burn, so he removed the air from his light bulb, creating a vacuum, around the filament. Then the search for the proper filament began.

Figure 2
Many of Edison's greatest inventions, including the phonograph and the electric light bulb, were developed in his laboratory in Menlo Park, New Jersey. In fact, Edison was called "The Wizard of Menlo Park."

Experimentation and Improvement

Edison unsuccessfully experimented with more than 1,600 materials, including plant fibers, fishing line, hair, and platinum. Then, Edison and his team experimented with carbon, a non-metallic element that was inexpensive and glowed when current was passed through it. Because carbon can't be shaped into a wire, Edison had to coat other substances with carbon to make the light bulb filament. In 1879, one of Edison's researchers tested a thin piece of cotton thread coated with carbon as a filament. The tiny filament glowed for 15.5 hours before Edison increased the voltage and it burned out. In this way, Edison used experimentation as a method toward new development and patented his light bulb in 1880.

Lewis Latimer, an African-American inventor, also used experimentation to make significant improvements to the light bulb. He developed and patented a method for connecting the electrical wires and the carbon filament together in the base of the bulb in 1881 and a process to make a long-lasting carbon filament in 1882. Experimentation and improvements to electrical lighting continue today and longer-lasting light bulbs are the result.

Figure 3
Edison designed an airless glass bulb in which to test filament materials.

Figure 4
Lewis Latimer significantly improved the carbon filament, making electric light bulbs more efficient and durable.

Figure 5
Because of continued experimentation and improvements, modern incandescent light bulbs, like those that help light this city, last an average of 875 hours. Some specially designed bulbs last as long as 20,000 hours.

The Study of Matter and Energy

Edison and Latimer, like all scientists, attempted to answer questions by performing tests and recording the results. When you answer a question or solve a problem by conducting a test, you are taking the scientific approach.

Experiments with electricity and light are part of physical science, the study of matter and energy. Two of the main branches of physical science are chemistry and physics. Chemistry is the study of what substances are made of and how they change. Physics is the study of matter and energy, including light and sound.

Steps for Experiments

1. Limit Independent Variables.
Only one independent variable should be used in any experiment.

2. Use a control.
There must be a sample group that is treated like the others except the independent variable isn't applied.

3. Repeat the experiment.
To insure that the results are valid, experiments must be repeated several times.

Experimentation

Experiments must be carefully planned in order to insure the accuracy of the results. Scientists begin by defining what they expect the experiment to prove. Edison's filament experiments were designed to find which material would act as the best filament for an incandescent light bulb. Edison tested filament materials by placing them in airless bulbs and then running electric current through them.

Variables and Controls in an Experiment

When scientists conduct experiments, they must make sure that only one factor affects the results of the experiment. The factor being changed is called the independent variable. The dependent variable is what is measured or observed to obtain the results of the experiment. In Edison's filament experiment, the independent variables were the different materials that were tested as filaments. The dependent variable was how each of the tested substances reacted when electric current flowed through them.

The conditions that stay the same in an experiment are called constants. The constants in Edison's filament experiments

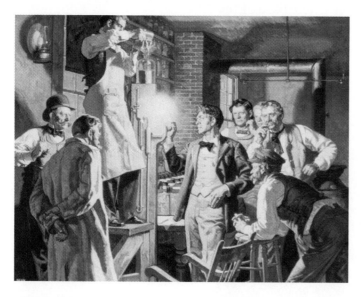

Figure 6
In this illustration, Edison (third from left) tests the electric light as his fellow researchers observe the results.

included the voltage applied and using the same type of bulb to surround each filament.

Edison changed a factor that should have been a constant, however, when he increased the voltage running through the carbonized cotton thread. Well-planned experiments also need a control—a sample that is treated like all the others except the independent variable isn't applied.

Interpreting Data

The observations and measurements that a scientist makes in an experiment are called data. Data must be carefully studied before questions can be answered or problems can be solved. Scientists repeat their experiments many times to make sure that their results are accurate.

Drawing Conclusions, Eliminating Biases

A conclusion is a statement that summarizes the results of the data that is obtained by the experiment. It is important that scientists are not influenced or biased by what they think the results will be or by what they want the results to be. A bias is a prejudice or an opinion. To avoid a biased conclusion it is important that scientists look at their data carefully and make sure their conclusion is based on their data. If more than one conclusion is possible, scientists often will conduct more tests to eliminate some of the possibilities or to find the best solution. Edison found several materials that glowed when a voltage was applied, but they were not suitable for lighting for various reasons. He found that carbon glowed when a voltage was applied and it had other qualities that made it a good choice for the filament. However, since carbon was brittle and did not form a wire, he had to keep experimenting to find the best material to support the carbon to make the filament.

"Results? Why, man, I have gotten lots of results! If I find 10,000 ways something won't work, I haven't failed. I am not discouraged, because every wrong attempt discarded is another step forward. Just because something doesn't do what you planned it to do doesn't mean it's useless.... Reverses should prove an incentive to great accomplishment.... There are no rules here, we're just trying to accomplish something."
-Thomas Edison

Figure 7
This quote from Thomas Edison is an example of a conclusion.

You Do It

Thomas Edison is only one of many inventors who conducted numerous experiments before creating a successful invention. Research the experiments that went into the invention of the telephone. How long did it take? How is the technology of the telephone that was used in 1900 different from the phone many people use today?

Waves

On a breezy day in Maui, Hawaii, windsurfers ride the ocean waves. What forces are operating on the windsurfer and his sailboard? The wind catches the sails and helps propel the sailboard, but other forces also are at work—waves. Waves carry energy. You can see the ocean waves in this picture, but there are many kinds of waves you cannot see. Microwaves heat your food, radio waves transmit the music you listen to into your home, and sound waves carry that music from the radio to your ears. In this chapter, you will learn about different types of waves and how they behave.

What do you think?

Science Journal What is this picture about? What causes the light and dark areas? Hint: *Some sunglasses have this kind of lens.* Write your answer or best guess in your Science Journal.

It's a beautiful autumn day. You are sitting by a pond in a park. Music blares from a school marching band practicing for a big game. The music is carried by waves. A fish jumps, making a splash. Waves spread past a leaf that fell from a tree, causing the leaf to move. In the following activity, you'll observe how waves carry energy that can cause objects to move.

Observe wave behavior

1. Fill a large, clear plastic plate with 1 cm of water.
2. Use a dropper to release a single drop of water onto the water's surface. Repeat.
3. Float a cork or straw on the water.
4. When the water is still, repeat step 2 from a height of 10 cm, then again from 20 cm.

Observe

In your Science Journal, record your observations and describe the movements of the floating object.

Before You Read

FOLDABLES
Reading & Study Skills

Making a Concept Map Study Fold Make the following Foldable to organize information by diagramming ideas about waves.

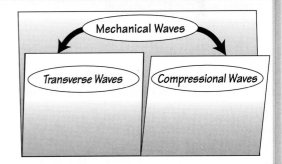

1. Place a sheet of paper in front of you so the long side is at the top. Fold the bottom of the paper to the top, stopping about four centimeters from the top.
2. Draw an oval above the fold. Write *Mechanical Waves* inside the oval.
3. Fold the paper in half from the left side to the right side and then unfold. Through the top thickness of the paper, cut along the fold line to form two tabs.
4. Draw an oval on each tab. Write *Transverse Waves* in one oval and *Compressional Waves* in the other, as shown. Draw arrows from the large oval to the smaller ovals.
5. As you read the chapter, write information about the two types of mechanical waves under the tabs.

SECTION

1

What are waves?

As You Read

What You'll Learn

- **Explain** the relationship among waves, energy, and matter.
- **Describe** the difference between transverse waves and compressional waves.

Vocabulary

wave
mechanical wave
transverse wave
compressional wave
electromagnetic wave

Why It's Important

You can hear music and other sounds because of waves.

Figure 1
The wave and the thrown ball carry energy.

What is a wave?

When you are relaxing on an air mattress in a pool and someone does a cannonball off the diving board, you suddenly find yourself bobbing up and down. You can make something move by giving it a push or pull, but the person jumping didn't touch your air mattress. How did the energy from the cannonball dive travel through the water and move your air mattress? The up-and-down motion was caused by the peaks and valleys of the ripples that moved from where the splash occurred. These peaks and valleys make up water waves.

Waves Carry Energy **Waves** are rhythmic disturbances that carry energy without carrying matter, as shown in **Figure 1A.** You can see the energy of the wave from the speedboat traveling outward, but the water only moves up and down. If you've ever felt a clap of thunder, you know that sound waves can carry large amounts of energy. You also transfer energy when you throw something to a friend, as in **Figure 1B.** However, there is a difference between a moving ball and a moving wave. A ball is made of matter, and when it is thrown, the matter moves from one place to another. So, unlike the moving wave, throwing a ball involves the transport of matter as well as energy.

A The waves created by a boat move mostly up and down, but the energy travels outward from the boat.

B When the ball is thrown, the ball, as well as the energy put into the throw, moves forward.

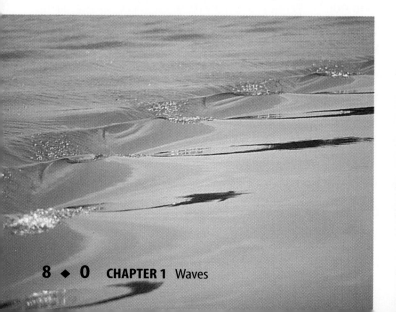

Figure 2

A As the students pass the ball, the students' positions do not change—only the position of the ball changes. B The molecules in water bump each other and pass the energy of a wave, even though they don't travel with the wave.

A Model for Waves

How does a wave carry energy without transporting matter? Imagine a line of people, as shown in **Figure 2A.** The first person in line passes a ball to the second person, who passes the ball to the next person, and so on. Passing a ball down a line of people is a model for how waves can transport energy without transporting matter. Even though the ball has traveled, the people in line have not moved. In this model, you can think of the ball as representing energy. What do the people in line represent?

Think about the ripples on the surface of a pond. The energy carried by the ripples travels through the water. The water is made up of water molecules. It is the individual molecules of water that pass the wave energy, just as the people in **Figure 2A** pass the ball. The water molecules transport the energy in a water wave by colliding with the molecules around them, as shown in **Figure 2B.**

☑ **Reading Check** *What is carried by waves?*

Mechanical Waves

In the wave model, the ball could not be transferred if the line of people didn't exist. The wave energy of a water wave could not be transferred if no water molecules existed. These types of waves, which use matter to transfer energy, are called **mechanical waves.** The matter through which a mechanical wave travels is called a medium. For ripples on a pond, the medium is the water.

A mechanical wave travels as energy is transferred from particle to particle in the medium. For example, a sound wave is a mechanical wave that can travel through air, as well as solids, liquids, and other gases. The sound wave travels through air by transferring energy from gas molecule to gas molecule. Without a medium such as air, you would not hear sounds. In outer space sound waves can't travel because there is no air.

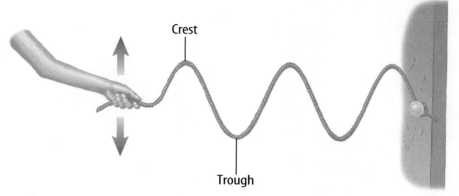

Crest

Trough

Figure 3
The high points on the wave are called crests and the low points are called troughs.

Transverse Waves In a mechanical **transverse wave,** the wave energy causes the matter in the medium to move up and down or back and forth at right angles to the direction the wave travels. You can make a model of a transverse wave. Stretch a long rope out on the ground. Hold one end in your hand. Now shake the end in your hand back and forth. By adjusting the way you shake the rope, you can create a wave that seems to slide along the rope.

When you first started shaking the rope, it might have appeared that the rope itself was moving away from you. But it was only the wave that was moving away from your hand. The wave energy moves through the rope, but the matter in the rope doesn't travel. You can see that the wave has peaks and valleys at regular intervals. As shown in **Figure 3,** the high points of transverse waves are called crests. The low points are called troughs.

✔ Reading Check *What are the highest points of transverse waves called?*

Figure 4
A compressional wave can travel through a coiled spring toy.

A As the wave motion begins, the coils near the string are close together and the other coils are far apart.

B The wave, seen in the squeezed and stretched coils, travels along the spring.

C The string and coils did not travel with the wave. Each coil moved only slightly forward and then back to its original position.

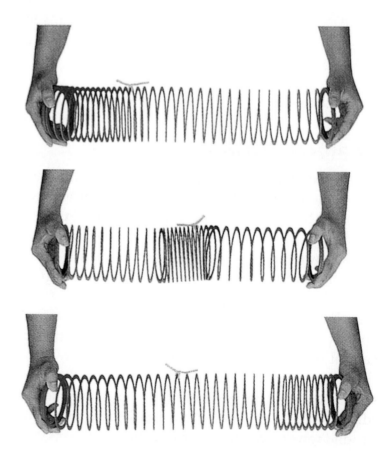

Compressional Waves Mechanical waves can be either transverse or compressional. In a **compressional wave,** matter in the medium moves forward and backward in the same direction that the wave travels. You can make a compressional wave by squeezing together and releasing several coils of a coiled spring toy, as shown in **Figure 4.**

You see that the coils move only as the wave passes. They then return to their original position. So, like transverse waves, compressional waves carry only energy forward along the spring. In this example, the spring is the medium the wave moves through, but the spring does not move along with the wave.

Sound Waves Sound waves are compressional waves. How do you make sound waves when you talk or sing? If you hold your fingers against your throat while you hum, you can feel vibrations. These vibrations are the movements of your vocal cords. If you touch a stereo speaker while it's playing, you can feel it vibrating, too. All waves are produced by something that is vibrating.

Making Sound Waves

How do vibrating vocal cords, strings, and other objects make sound waves? To find out, look at the drumhead stretched over the open end of the drum shown in **Figure 5.** When the drumhead moves upward, it touches some of the molecules that make up the air. When everything is quiet, these molecules are spaced about the same distance apart. However, when the drumhead moves upward, it pushes the molecules together. The group of molecules that are squeezed together is called a compression.

When the drumhead moves downward, the molecules have more room and move away from each other. A place where molecules are far apart is called a rarefaction (rar uh FAK shun). These disturbed molecules then collide with the molecules next to them, transferring the energy they are carrying. This causes the compression and the rarefaction to move away from the drumhead.

Figure 5
A vibrating drumhead makes compressions and rarefactions in the air. *How do your vocal cords make compressions and rarefactions in air?*

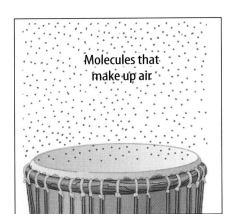

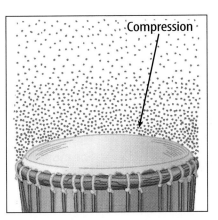

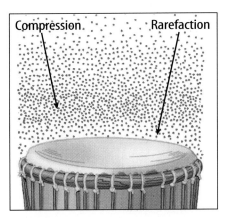

Electromagnetic Waves

When you listen to the radio, watch TV, or use a microwave oven to cook, you use a different kind of wave—one that doesn't need matter as a medium.

Waves that do not require matter to carry energy are called **electromagnetic waves.** Electromagnetic waves are transverse waves that are produced by the motion of electrically charged particles. Just like mechanical waves, electromagnetic waves can travel through a medium such as a solid, liquid, or gas. Radio waves are electromagnetic waves that travel through the air from a radio station, and then through the solid walls of your house to reach your radio. However, unlike mechanical waves, electromagnetic waves can travel through outer space or through a vacuum where no matter exists.

Useful Waves In space, which has no air or any other medium, orbiting satellites beam radio waves to TVs, radios, and cellular phones on Earth's surface. However, radio waves are not the only electromagnetic waves traveling in space. You use sunscreen to protect yourself from ultraviolet rays. Infrared and ultraviolet waves travel from the Sun through space before they reach Earth's atmosphere. Infrared waves feel warm when they strike your skin. Other useful electromagnetic waves include X rays and visible light. X rays are useful not only in medical applications, but also for security checks in airports as luggage is scanned. And without visible light you wouldn't see color or be able to read this page.

Physics
INTEGRATION

Maybe you've used a global positioning system (GPS) receiver to determine your location while driving, boating, or hiking. Earth-orbiting satellites send electromagnetic radio waves that transmit their exact locations and times of transmission. The GPS receiver uses information from four of these satellites to determine your location to within about 16 m.

Section ① Assessment

1. Describe the movement of a floating object on a pond when struck by a wave.

2. Why can't a sound wave travel from a satellite to Earth?

3. Give one example of a transverse wave and one example of a compressional wave. How are they similar and different?

4. What is the difference between a mechanical wave and an electromagnetic wave?

5. **Think Critically** How is it possible for a sound wave to transmit energy but not matter?

Skill Builder Activities

6. **Concept Mapping** Create a concept map that shows the relationships among the following: *waves, mechanical waves, electromagnetic waves, compressional waves,* and *transverse waves.* **For more help, refer to the** Science Skill Handbook.

7. **Using a Word Processor** Use word-processing software to write short descriptions of the waves you encounter during a typical day. **For more help, refer to the** Technology Skill Handbook.

Wave Properties

Amplitude

Can you describe a wave? One way might be to tell how high a water wave rises above, or falls below, the normal level. This distance is called the wave's amplitude. The **amplitude** of a transverse wave is one-half the distance between a crest and a trough, as shown in **Figure 6A.** In a compressional wave, the amplitude is greater when the particles of the medium are squeezed closer together in each compression and spread farther apart in each rarefaction.

Amplitude and Energy A wave's amplitude is related to the energy that the wave carries. For example, the electromagnetic waves that make up bright light have greater amplitudes than the waves that make up dim light. Waves of bright light carry more energy than the waves that make up dim light. In a similar way, loud sound waves have greater amplitudes than soft sound waves. Loud sounds carry more energy than soft sounds. If a sound is loud enough, it can carry enough energy to damage your hearing.

As you can see in **Figure 6B,** when a hurricane strikes a coastal area, the resulting water waves can damage anything that stands in their path. The waves caused by a hurricane carry more energy than the small waves or ripples on a pond.

Figure 6
A transverse wave has an amplitude.

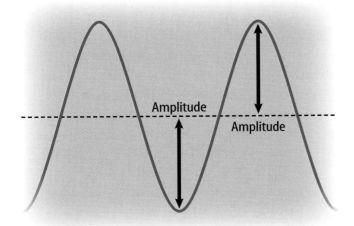

A The amplitude is a measure of how high the crests are or how deep the troughs are.

B A wave of large amplitude is responsible for this damage.

A For transverse waves, measure crest to crest or trough to trough.

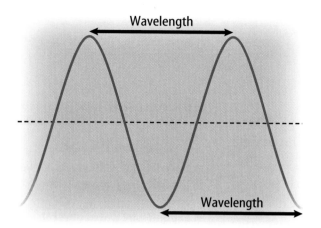

B For compressional waves, measure from compression to compression or from rarefaction to rarefaction

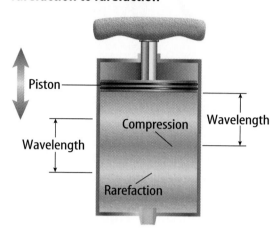

Figure 7
Wavelength is measured differently for transverse and compressional waves.

Figure 8
The wavelengths and frequencies of electromagnetic waves vary.

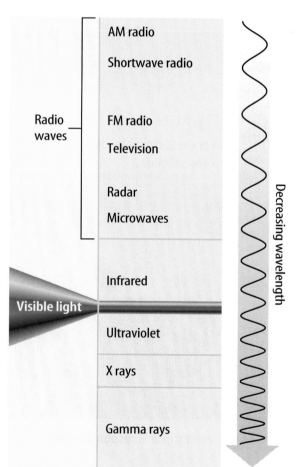

Earth Science INTEGRATION

The devastating effect that a wave with large amplitude can have is seen in the aftermath of tsunamis. Tsunamis are huge sea waves that are caused by underwater earthquakes along faults on the seafloor. The movement of the seafloor along the fault produces the wave. As the wave moves toward shallow water and slows down, the amplitude of the wave grows. The tremendous amounts of energy tsunamis carry cause great damage when they move ashore.

Wavelength

Another way to describe a wave is by its wavelength. For a transverse wave, **wavelength** is the distance from the top of one crest to the top of the next crest, or from the bottom of one trough to the bottom of the next trough, as shown in **Figure 7A.** For a compressional wave, the wavelength is the distance between the center of one compression and the center of the next compression, or from the center of one rarefaction to the center of the next rarefaction, as shown in **Figure 7B.**

Wavelength is an important characteristic of a wave. For example, the difference between red light and green light is that they have different wavelengths. The wavelength of red light is longer than the wavelength of green light. It is the wavelength of visible light that determines its color. Some electromagnetic waves, like X rays, have extremely short wavelengths. Others, like microwaves, have longer wavelengths. The range of wavelengths of electromagnetic waves is shown in **Figure 8.**

Frequency

The **frequency** of a wave is the number of wavelengths that pass a given point in 1s. The unit of frequency is the number of wavelengths per second, or hertz (Hz). Recall that waves are produced by something that vibrates. The faster the vibration is, the higher the frequency is of the wave that is produced.

✔ Reading Check *How is the frequency of a wave measured?*

A Sidewalk Model For waves that travel with the same speed, frequency and wavelength are related. To model this relationship, imagine people on two parallel moving sidewalks in an airport, as shown in **Figure 9.** One sidewalk has four travelers spaced 4 m apart. The other sidewalk has 16 travelers spaced 1 m apart.

Now imagine that both sidewalks are moving at the same speed and approaching a pillar between them. On which sidewalk will more people go past the pillar? On the sidewalk with the shorter distance between people, four people will pass the pillar for each one person on the other sidewalk. When four people pass the pillar on the first sidewalk, 16 people pass the pillar on the second sidewalk.

Figure 9
This moving sidewalk illustration shows how wavelength and frequency are related.

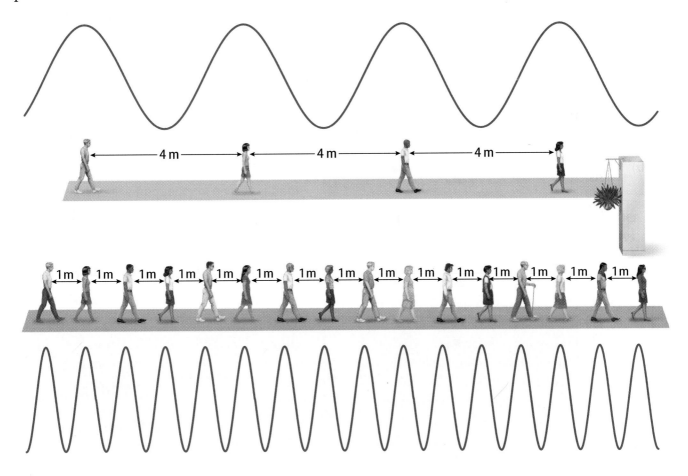

Activity

Waves on a Spring

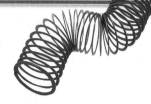

Waves are rhythmic disturbances that carry energy through matter or space. Studying waves can help you understand how the Sun's energy reaches Earth and sounds travel through the air.

What You'll Investigate
What are some of the properties of transverse and compressional waves on a coiled spring?

Materials
long, coiled spring toy
colored yarn (5 cm)
meterstick
stopwatch

Goals
- **Create** transverse and compressional waves on a coiled spring toy.
- **Investigate** wave properties such as speed and amplitude.

Safety Precautions 🥽
WARNING: *Avoid overstretching or tangling the spring to prevent injury or damage.*

Procedure

1. **Prepare** a data table such as the one shown.

Wave Data	
Length of stretched spring toy	
Average time for a wave to travel from end to end—step 4	
Average time for a wave to travel from end to end—step 5	

2. Work in pairs or groups and clear a place on an uncarpeted floor about 6 m × 2 m.

3. Stretch the springs between two people to the length suggested by your teacher. Measure the length.

4. Create a wave with a quick, sideways snap of the wrist. Time several waves as they travel the length of the spring. Record the average time in your data table.

5. Repeat step 4 using waves that have slightly larger amplitudes.

6. Squeeze together about 20 of the coils. Observe what happens to the unsqueezed coils. Release the coils and observe.

7. Quickly push the spring toward your partner, then pull it back.

8. Tie the yarn to a coil near the middle of the spring. Repeat step 7, observing the string.

Conclude and Apply

1. **Classify** the wave pulses you created in each step as compressional or transverse.
2. **Calculate** and compare the speeds of the waves in steps 4 and 5.
3. **Classify** the unsqueezed coils in step 6 as a compression or a rarefaction.
4. **Compare and contrast** the motion of the yarn with the motion of the wave.

𝒞ommunicating Your Data

Write a summary paragraph of how this activity demonstrated any of the vocabulary words from the first two sections of the chapter. **For more help,** refer to the Science Skill Handbook.

Wave Behavior

Reflection

What causes the echo when you yell across an empty gymnasium or down a long, empty hallway? Why can you see your face when you look in a mirror? The echo of your voice and the face you see in the mirror are caused by wave reflection.

Reflection occurs when a wave strikes an object or surface and bounces off. An echo is reflected sound. Sound reflects from all surfaces. Your echo bounces off the walls, floor, ceiling, furniture, and people. You see your face in a mirror or a still pond, as shown in **Figure 11A,** because of reflection. Light waves produced by a source of light such as the Sun or a lightbulb bounce off your face, strike the mirror, and reflect back to your eyes.

A mirror is smooth and even, therefore you see a clearly reflected image. However, when light reflects from an uneven or rough surface, you can't see an image because the reflected light scatters in many different directions, as shown in **Figure 11B.**

✔ **Reading Check** *What causes reflection?*

As You Read

What You'll Learn
- **Explain** how waves can reflect from some surfaces.
- **Explain** how waves change direction when they move from one material into another.
- **Describe** how waves are able to bend around barriers.

Vocabulary
reflection
refraction
diffraction
interference

Why It's Important
The reflection of waves enables you to see objects around you.

A The smooth surface of a still pond allows light to reflect back to you so you can see your image.

B The rough, uneven surface of the same pond causes the light to reflect in many directions so that no sharp image is visible.

Figure 11
The image formed by reflection depends on the smoothness of the surface.

Refraction

You've already seen that a wave changes direction when it reflects from a surface. Waves also can change direction in another way. Perhaps you have tried to grab a sinking object when you are in a swimming pool, only to come up empty-handed. Yet you were sure you grabbed right where you saw the object. This happens because the light rays from the object change direction as they pass from the water into the air. The bending of a wave as it moves from one medium into another is called **refraction.**

Refraction and Wave Speed Remember that the speed of a wave can be different in different materials. For example, light waves travel faster in air than in water. Refraction occurs when the speed of a wave changes as it passes from one substance to another, as shown in **Figure 12.** A line that is perpendicular to the water's surface is called the normal. When a light ray passes from air into water, it slows down and bends toward the normal. When the ray passes from water into air, it speeds up and bends away from the normal. The larger the change in speed of the light wave is, the larger the change in direction is.

You notice refraction when you look down into a fishbowl. Refraction makes the fish appear to be closer to the surface but farther away from you than it is, as shown in **Figure 13.** Light rays reflected from the fish are bent away from the normal as they pass from water to air. Your brain interprets the light that enters your eyes by assuming that light rays always travel in straight lines. As a result, the light rays seem to be coming from a fish that is in a different location.

Figure 12
A wave is refracted when it changes speed. **A** As the light ray passes from air to water, it refracts toward the normal. **B** As the light ray passes from water to air, it refracts away from the normal.

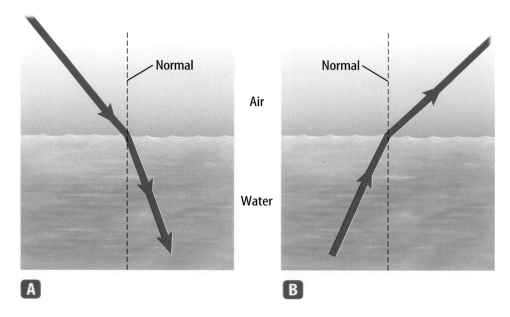

Color from Refraction Refraction causes prisms to separate sunlight into many colors and produces rainbows too. **Figure 14** illustrates how refraction and reflection produce a rainbow when light waves from the Sun pass into and out of water droplets in the air.

✔ **Reading Check** *What produces a rainbow?*

Diffraction

Why can you hear music from the band room when you are down the hall? You can hear the music because the sound waves bend as they pass through an open doorway. This bending isn't caused by refraction. Remember that refraction occurs when waves change speed, but sound waves have the same speed in the band room and in the hallway. Instead the bending is caused by diffraction. **Diffraction** is the bending of waves around a barrier.

Diffraction of Light Waves Can light waves diffract, too? You can hear your friends in the band room but you can't see them until you reach the open door. Therefore, you know the light waves do not diffract as much as sound waves do.

Are light waves able to diffract at all? Light waves do bend around the edges of an open door. However, for an opening as wide as a door, the amount the light bends is extremely small. As a result, the diffraction of light is far too small to allow you to see around a corner.

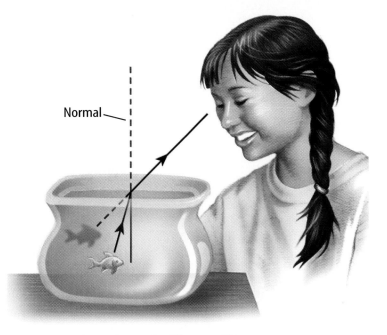

Normal

Figure 13
When you look at the goldfish in the water, the fish is in a different position than it appears.

Figure 14
Light rays refract as they enter and leave each water drop. Each color refracts at different angles because of their different wavelengths, so they separate into the colors of the spectrum.

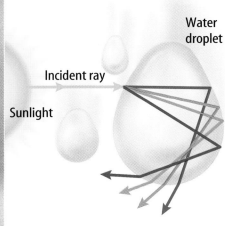

Water droplet

Incident ray

Sunlight

Diffraction and Wavelength The reason that light waves don't diffract much when they pass through an open door is that the wavelengths of visible light are much smaller than the width of the door. Light waves have wavelengths between 400 and 700 billionths of a meter, while the width of doorway is about one meter. Sound waves that you can hear have much longer wavelengths. They bend more easily around the corners of an open door. A wave is diffracted more when its wavelength is similar in size to the barrier or opening.

Diffraction of Water Waves Perhaps you have noticed water waves bending around barriers. For example, when water waves strike obstacles such as the islands shown in **Figure 15,** they don't stop moving. Here the size and spacing of the islands is not too different from the wavelength of the water waves. So the water waves bend around the islands, and keep on moving. They also spread out after they pass through openings between the islands. If the islands were much larger than the water wavelength, less diffraction would occur.

What happens when waves meet?

Suppose you throw two pebbles into a still pond. Ripples spread from the impact of each pebble and travel toward each other. What happens when two of these ripples meet? Do they collide like billiard balls and change direction? Waves behave differently from billiard balls when they meet. In fact, after they pass, waves continue moving as though the other waves never existed.

Figure 15
Water waves bend or diffract around these islands. More diffraction occurs when the object is closer in size to the wavelength.

Wave Interference When two waves overlap a new wave is formed by adding the two waves together. The ability of two waves to combine and form a new wave when they overlap is called **interference.** After they overlap, the individual waves continue to travel on in their original form.

The different ways waves can interfere are shown in **Figure 16** on the next page. Sometimes when the waves meet, the crest of one wave overlaps the crest of another wave. This is called constructive interference. The amplitudes of these combining waves add together to make a larger wave while they overlap. Destructive interference occurs when the crest of one wave overlaps the trough of another wave. In destructive interference, the amplitudes of the two waves combine to make a wave with a smaller amplitude. If the two waves have equal amplitudes and meet crest to trough, they cancel each other during the overlap.

Waves and Particles Like waves of water, when light travels through a small opening, such as a narrow slit, the light spreads out in all directions on the other side of the slit. If small particles, instead of waves, were sent through the slit, they would continue in a straight line without spreading. The spreading, or diffraction, is only a property of waves. Interference also doesn't occur with particles. If waves meet, they reinforce or cancel each other, then travel on. If particles approach each other, they either collide and scatter or miss each other completely. Interference, like diffraction, is a property of waves, not particles.

SCIENCE *Online*

Research Visit the Glencoe Science Web site at **science.glencoe.com** for more information about wave interference.

Problem-Solving Activity

Can you create destructive interference?

Your brother is vacuuming and you can't hear the television. Is it possible to diminish the sound of the vacuum so you can hear the TV? Can you eliminate unpleasant sounds and keep the sounds you do want to hear?

Identifying the Problem

It is possible to create a frequency that will destructively interfere with the sound of the vacuum and not the television. The graph shows the waves created by the vacuum and the television.

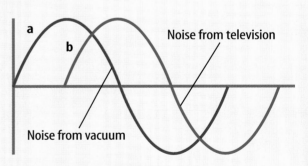

Solving the Problem

1. Can you create the graph of a wave that will eliminate the noise from the vacuum but not the television?
2. Can you create the graph of a wave that would amplify the sound of the television?

NATIONAL GEOGRAPHIC **VISUALIZING INTERFERENCE**

Figure 16

Whether they are ripples on a pond or huge ocean swells, when water waves meet they can combine to form new waves in a process called interference. As shown below, wave interference can be constructive or destructive.

Constructive Interference

In constructive interference, a wave with greater amplitude is formed.

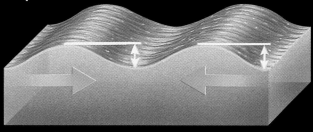

The crests of two waves—A and B—approach each other.

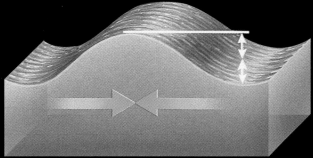

The two waves form a larger wave where the crests of both waves overlap.

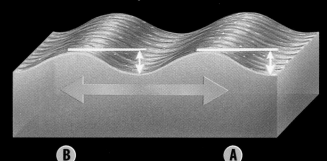

The original waves pass through each other and go on as they started.

Destructive Interference

In destructive interference, a wave with a smaller amplitude is formed.

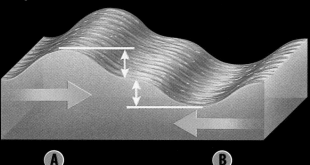

The crest of one wave approaches the trough of another.

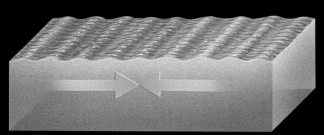

If the two waves have equal amplitude, they momentarily cancel when they meet.

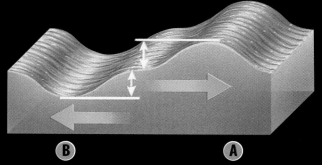

The original waves pass through each other and go on as they started.

Using Interference You might have seen someone use a power lawn mower or a chain saw. In the past, many people who performed these tasks damaged their hearing because of the loud noises produced by these machines. Today, specially designed ear protectors absorb the sound from lawn mowers and chain saws. The ear protectors lower the amplitudes of the harmful waves. The waves that reach the ears have smaller amplitudes and won't damage eardrums.

Reducing Noise Pilots and passengers of small planes have a more complicated problem. They can't use ear protectors to shut out all the noise of the plane's motor. If they did, the pilots wouldn't be able to hear instructions from air-traffic controllers, and the passengers wouldn't be able to hear each other talk. To solve this problem, engineers invented special devices that contain electronic circuits as shown in **Figure 17.** These circuits detect the vibrations from the aircraft that make noise and produce sound frequencies that destructively interfere with those vibrations. However, the sound frequencies produced do not interfere with human voices, so people can hear and understand normal conversation. In these examples, destructive interference can be a benefit.

Figure 17
Some airplane pilots use special ear protectors that interfere with engine noise but don't block human voices.

Section 3 Assessment

1. Why don't you see your reflection in a building made of rough, white stone?

2. If you're standing on one side of a building, how are you able to hear the siren of an ambulance on the other side?

3. What behavior of light enables magnifying glasses and contact lenses to bend light rays and help people see more clearly?

4. What is diffraction? How does the amount of diffraction depend on wavelength?

5. **Think Critically** Why don't light rays that stream through an open window into a darkened room spread evenly through the entire room?

Skill Builder Activities

6. **Comparing and Contrasting** When light rays pass from water into a certain type of glass, the rays refract toward the normal. Compare and contrast the speed of light in water and in the glass. **For more help, refer to the** Science Skill Handbook.

7. **Communicating** Watch carefully as you travel home from school or walk down your street. What examples of wave reflection and refraction do you notice? Describe each of these in your Science Journal and explain your reasons. **For more help, refer to the** Science Skill Handbook.

Activity

Design Your Own Experiment

Wave Speed

When an earthquake occurs, the waves of energy are recorded at points all over the world by instruments called seismographs. By comparing the data that they collected from their seismographs, scientists discovered that the interior of Earth must be made of layers of different materials. How did the seismographs tell them that Earth is not the same material all the way through?

Recognize the Problem

Can the speed of a wave be used to identify the medium through which it travels?

Form a Hypothesis

Think about what you know about the relationship between the frequency, wavelength, and speed of a wave in a medium. Make a hypothesis about how you can measure the speed of a wave within a medium, and use that information to identify an unknown medium.

Goals
- **Measure** the speed of a wave within a coiled spring toy.
- **Predict** whether the speed you measured will be different in other types of coiled spring toys.

Materials
coiled spring toy   meterstick
stopwatch   tape
clock with a second hand
Alternate materials

Safety Precautions

Test Your Hypothesis

Plan

1. Make a data table in your Science Journal like the one shown.

2. In your Science Journal, write a detailed description of the coiled spring toy you are going to use. Be sure to include its mass and diameter, the width of a coil, and what it is made of.

3. **Decide** as a group how you will measure the frequency and length of waves in the spring toy. What are your variables? Which variables must be controlled? What variable do you want to measure?

4. Repeat your experiment three times.

Wave Data	Trial 1	Trial 2	Trial 3
Length spring was stretched (m)			
Number of crests			
Wavelength (m)			
# of vibrations timed			
# of seconds vibrations were timed			
Wave speed (m/s)			

Do

1. Make sure your teacher approves your plan before you start.

2. Carry out the experiment.

3. While you are doing the experiment, record your observations and measurements in your data table.

Analyze Your Data

1. **Calculate** the frequency of the waves by dividing the number of vibrations you timed by the number of seconds you timed them. Record your results in your data table.

2. Use the following formula to calculate the speed of a wave in each trial.

$$\text{wavelength} \times \frac{\text{wave}}{\text{frequency}} = \frac{\text{wave}}{\text{speed}}$$

3. Average the wave speeds from your trials to determine the average speed of a wave in your coiled spring toy.

Draw Conclusions

1. How do the variables affect the wave speed in spring toys? Was your hypothesis supported?

2. Would it make a difference if an earthquake wave were transmitted through Earth's solid mantle or the molten outer core?

ommunicating Your Data

Post a description of your coiled spring toy and the results of your experiment on a bulletin board in your classroom. **Compare and contrast** your results with other students in your class.

Waves, Waves, and More Waves

Did you know...

...You are constantly surrounded by a sea of waves even when you're on dry land! Electromagnetic waves around us are used to cook our food and transmit signals to our radios and televisions. Light itself is an electromagnetic wave.

...The highest recorded ocean wave was 34 meters high, which is comparable to the height of a ten-story building. This super wave was seen in the North Pacific Ocean and recorded by the crew of the naval ship *USS Ramapo* in 1933.

...Tsunamis—huge ocean waves—travel at speeds near 966 km/h.

...Waves let dolphins see with their ears! A dolphin sends out ultrasonic pulses, or clicks, at speeds of 800 pulses per second. These sound waves echo back to the dolphin after they hit another object. This process—echolocation—allows dolphins to recognize obstacles and meals.

. . . Earthquakes are caused by a variety of seismic waves— waves of energy that ripple through Earth after sub-surface rock breaks suddenly. The fastest are P and S waves. P waves are compressional waves that travel at about 8 km/s. S waves, which move like ocean waves, travel at about 4.8 km/s.

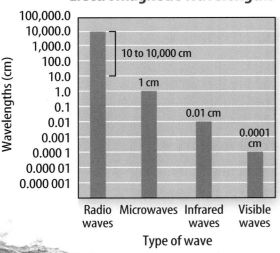

Electromagnetic Wavelengths

Wavelengths (cm)

100,000.0
10,000.0
1,000.0
100.0
10.0
1.0
0.1
0.01
0.001
0.000 1
0.000 01
0.000 001

10 to 10,000 cm

1 cm

0.01 cm

0.0001 cm

Radio waves | Microwaves | Infrared waves | Visible waves

Type of wave

. . . Radio waves from space were discovered in 1932 by Karl G. Jansky, an American engineer. His amazing discovery led to creation of radio astronomy, a field that explores parts of the universe that are hidden by interstellar dust or too distant for optical observation.

Do the Math

1. A museum with a dolphin exhibit plays dolphin clicks for its visitors at a speed 250 times slower than the speed at which the dolphins emit them. How many clicks do the visitors hear in 10 s?
2. Tsunamis form in the ocean when an earthquake occurs on the ocean floor. How long will it take a tsunami to travel 4,500 km?
3. Make a bar graph to show the speeds of P waves, S waves and tsunamis. Use km/h as your unit of speed.

Go Further

Go to **science.glencoe.com** to learn about discoveries by radio astronomers. Graph the distances of these discoveries from Earth.

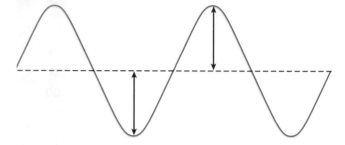

Reviewing Main Ideas

Section 1 What are waves?

1. Waves are rhythmic disturbances that carry energy but not matter.

2. Mechanical waves can travel only through matter. Electromagnetic waves can travel through matter and space.

3. In a mechanical transverse wave, matter in the medium moves back and forth at right angles to the direction the wave travels.

4. In a compressional wave, matter in the medium moves forward and backward in the same direction as the wave. *How does the boat in the picture move as the water wave goes by?*

Section 2 Wave Properties

1. The amplitude of a transverse wave is one half the distance between a crest and a trough.

2. The energy carried by a wave increases as the amplitude increases.

3. Wavelength is the distance between neighboring crests or neighboring troughs.

4. The frequency of a wave is the number of wavelengths that pass a given point in 1 s. *What characteristic of a wave is indicated in the figure at the top of the next column?*

5. Waves travel through different materials at different speeds.

Section 3 Wave Behavior

1. Reflection occurs when a wave strikes an object or surface and bounces off. *Why doesn't the foil show a clear image?*

2. The bending of a wave as it moves from one medium into another is called refraction. A wave changes direction, or refracts, when the speed of the wave changes.

3. The bending of waves around a barrier is called diffraction.

4. Interference occurs when two or more waves combine and form a new wave when they overlap.

FOLDABLES
Reading & Study Skills

After You Read

Use your Concept Map Study Fold to compare and contrast transverse and compressional mechanical waves.

Visualizing Main Ideas

Complete the following spider map about waves.

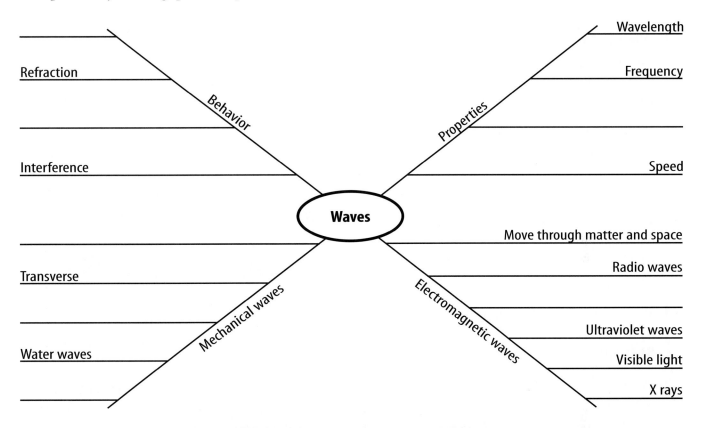

Wavelength

Frequency

Refraction

Behavior

Properties

Interference

Speed

Waves

Move through matter and space

Radio waves

Transverse

Mechanical waves

Electromagnetic waves

Ultraviolet waves

Water waves

Visible light

X rays

Vocabulary Review

Vocabulary Words

a. amplitude
b. compressional wave
c. diffraction
d. electromagnetic wave
e. frequency
f. interference
g. mechanical wave
h. reflection
i. refraction
j. transverse wave
k. wave
l. wavelength

Study Tip

After you've read a chapter, go back to the beginning and speed-read through what you've just read. This will help your memory.

Using Vocabulary

Using the list, replace the underlined words with the correct vocabulary words.

1. Diffraction is the change in direction of a wave when it strikes a surface.

2. The type of wave that has rarefactions is a transverse wave.

3. The distance between two adjacent crests of a transverse wave is the frequency.

4. The more energy a wave carries, the greater its wavelength is.

5. A mechanical wave can travel through space without a medium.

Chapter 1 Assessment

Checking Concepts

Choose the word or phrase that best answers the question.

1. What is the material through which mechanical waves travel?
   A) charged particles   C) a vacuum
   B) space   D) a medium

2. What is carried from particle to particle in a water wave?
   A) speed   C) energy
   B) amplitude   D) matter

3. What are the lowest points on a transverse wave called?
   A) crests   C) compressions
   B) troughs   D) rarefactions

4. What determines the pitch of a sound wave?
   A) amplitude   C) speed
   B) frequency   D) refraction

5. What is the distance between adjacent wave compressions?
   A) one wavelength   C) 1 m/s
   B) 1 km   D) 1 Hz

6. What occurs when a wave strikes an object or surface and bounces off?
   A) diffraction
   B) refraction
   C) a change in speed
   D) reflection

7. What is the name for a change in the direction of a wave when it passes from one medium into another?
   A) refraction   C) reflection
   B) interference   D) diffraction

8. What type of wave is a sound wave?
   A) transverse   C) compressional
   B) electromagnetic   D) refracted

9. When two waves overlap and interfere destructively, what does the resulting wave have?
   A) a greater amplitude
   B) more energy
   C) a change in frequency
   D) a lower amplitude

10. What is the difference between blue light and green light?
    A) They have different wavelengths.
    B) One is a transverse wave and the other is not.
    C) They have different pitch.
    D) One is mechanical and the other is not.

Thinking Critically

11. Explain what kind of wave—transverse or compressional—is produced when an engine bumps into a string of coupled rail-road cars on a track.

12. Is it possible for an electromagnetic wave to travel through a vacuum? Through matter? Explain your answers.

13. Why does the frequency of a wave decrease as the wavelength increases?

14. Why don't you see your reflected image when you look at a white, rough surface?

15. If a cannon fires at a great distance from you, why do you see the flash before you hear the sound?

Developing Skills

16. **Solving One-Step Equations** A microwave travels at the speed of light and has a wavelength of 0.022 m. If the wave speed is equal to the wavelength times the frequency, what is the frequency of the microwave?

17. Forming Hypotheses Form a hypothesis that can explain this observation. Waves A and B travel away from Earth through Earth's atmosphere. Wave A continues on into space, but wave B does not.

18. Recognizing Cause and Effect Explain how the object shown below causes compressions and rarefactions as it vibrates in air.

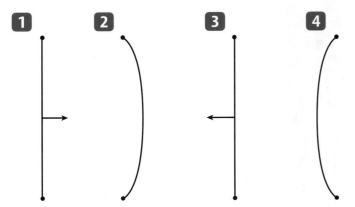

19. Comparing and Contrasting AM radio waves have wavelengths between about 200 m and 600 m, while FM radio waves have wavelengths of about 3 m. Why can AM radio signals often be heard behind buildings and mountains but FM radio signals cannot?

Performance Assessment

20. Making Flashcards Work with a partner to make flashcards for the bold-faced terms in the chapter. Illustrate each term on the front of the cards. Write the term and its definition on the back of the card. Use the cards to review the terms with another team.

TECHNOLOGY

Go to the Glencoe Science Web site at **science.glencoe.com** or use the **Glencoe Science CD-ROM** for additional chapter assessment.

Test Practice

Kamisha's science teacher told her that her remote control sent signals to the TV and VCR by using infrared waves. She decided to do some research about waves. The information she gathered is shown in the diagram below.

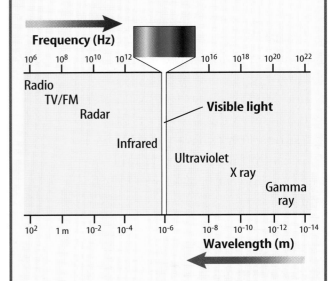

1. According to the diagram, which type of wave has a wavelength greater than 1 m?
A) radio
B) infrared
C) ultraviolet
D) X ray

2. According to the diagram, which type of wave has the HIGHEST frequency?
F) radio
G) ultraviolet
H) X ray
J) gamma ray

Sound

Have you ever experienced complete silence? Unless you have stood in a room like this one, you probably have not. This room is lined with materials that absorb sound waves and eliminate sound reflections. The sounds that you hear are created by vibrations. How do vibrations make sounds with different pitches? What makes a sound loud or quiet? In this chapter, you will learn the answers to these questions. You will also learn how musical instruments create sound and how the ear enables you to hear sound.

What do you think?

Science Journal Look at the picture below with a classmate. Discuss what might be happening. Here's a hint: *Sound is caused by vibrations.* Write your answer or best guess in your Science Journal.

EXPLORE ACTIVITY

When you speak or sing, you push air from your lungs past your vocal cords, which are two flaps of tissue inside your throat. When you tighten your vocal cords, you can make the sound have a higher tone. Do this activity to explore how you change the shape of your throat to vary the tone of sound.

Observe throat vibrations

1. Hold your fingers against the front of your throat and say *Aaaah.* Notice the vibration against your fingers.

2. Now vary the tone of this sound from low to high and back again. How do the vibrations in your throat change? Record your observations.

3. Change the sound to an *Ooooh.* What do you notice as you listen? Record your observations.

Observe

In your Science Journal, describe how the shape of your throat changed the tone.

Before You Read

FOLDABLES
Reading & Study Skills

Making a Question Study Fold Asking yourself questions helps you stay focused so you will better understand sound when you are reading the chapter.

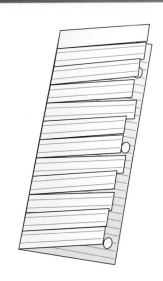

1. Place a sheet of notebook paper in front of you so the short side is at the top and the holes are on the right side. Fold the paper in half from the left side to the right side.

2. Through the top thickness of paper, cut along every third line from the outside edge to the fold, forming tabs.

3. Before you read the chapter, write a question you have about sound on the front of each tab. As you read the chapter, answer your questions and add more information.

What is sound?

As You Read

What **You'll Learn**

■ **Identify** the characteristics of sound waves.
■ **Explain** how sound travels.
■ **Describe** the Doppler effect.

Vocabulary
loudness
pitch
echo
Doppler effect

Why **It's Important**
Sound gives important information about the world around you.

Sound and Vibration

Think of all the sounds you've heard since you awoke this morning. Did you hear your alarm clock blaring, car horns honking, or locker doors slamming? Every sound has something in common with every other sound. Each is produced by something that vibrates.

Sound Waves

How does an object that is vibrating produce sound? When you speak, the vocal cords in your throat vibrate. These vibrations cause other people to hear your voice. The vibrations produce sound waves that travel to their ears. The other person's ears interpret these sound waves.

A wave carries energy from one place to another without transferring matter. An object that is vibrating in air, such as your vocal cords, produces a sound wave. The vibrating object causes air molecules to move back and forth. As these air molecules collide with those nearby, they cause other air molecules to move back and forth, transferring the energy of the sound wave. A sound wave is a compressional wave, like the wave moving through the coiled spring toy in **Figure 1.** In a compressional wave, particles in the material move back and forth along the direction the wave is moving. In a sound wave, air molecules move back and forth along the direction the sound wave is moving.

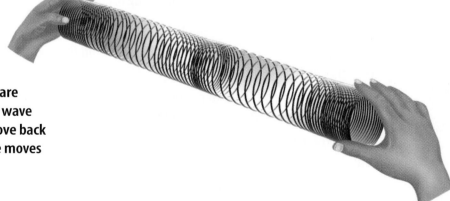

Figure 1
When the coils of a coiled spring toy are squeezed together, a compressional wave moves along the spring. The coils move back and forth as the compressional wave moves past them.

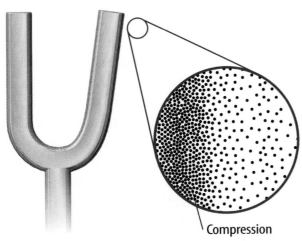

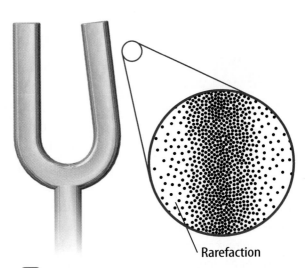

A When the tuning fork vibrates outward, it forces the air molecules next to it together, creating a region of compression.

B When the tuning fork moves back, the air molecules next to it spread apart, creating a region of rarefaction.

Making Sound Waves When an object vibrates, it exerts a force on the surrounding air. For example, as the end of the tuning fork moves outward into the air, it pushes the air molecules together, as shown in **Figure 2A.** As a result, a region where the air molecules are closer together, or more dense, is created. This region of higher density is called a compression. When the end of the tuning fork moves back, it creates a region of lower density called a rarefaction, as shown in **Figure 2B.** As the tuning fork continues to vibrate, a series of compressions and rarefactions is formed. The compressions and rarefactions move away from the tuning fork as molecules in these regions collide with other nearby molecules.

Like other waves, a sound wave can be described by its wavelength and frequency. The wavelength of a sound wave is shown in **Figure 3.** The frequency of a sound wave is the number of compressions or rarefactions that pass by a given point in one second. An object that vibrates faster forms a sound wave with a higher frequency.

Figure 2
A tuning fork makes a sound wave as the ends of the fork vibrate in the air. *Can a sound wave travel in a vacuum?*

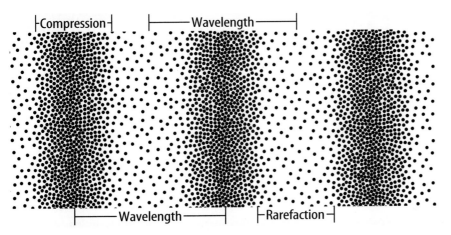

Figure 3
Wavelength is the distance from one compression to another or one rarefaction to another. *How does the wavelength of a sound wave relate to its frequency?*

TRY AT HOME
Mini LAB

Comparing and Contrasting Sounds

Procedure

1. Shake a set of **keys** and listen to the sound they make in air. Then submerge the keys and one ear in **water**. (A **tub** or a wide, deep **bowl** will work.) Again, shake the keys and listen to the sound. Use a **towel** to dry the keys.

2. Tie a **metal spoon** in the middle of a length of **cotton string**. Strike the spoon on something to hear it ring. Now press the ends of the string against your ears and repeat the experiment. What do you hear?

Analysis

1. Did you hear sounds transmitted through water and through string? Describe the sounds.

2. Compare and contrast the sounds in water and in air.

The Speed of Sound

Sound waves can travel through other materials besides air. Even though sound waves travel as compressions and rarefactions through different materials, they might travel at different speeds. As a sound wave travels through a material, the particles in the material it is moving through collide with each other. In a solid, molecules are closer together than in liquids or gases, so collisions between molecules occur more frequently than in liquids or gases. As a result, the speed of sound is usually fastest in solids, where molecules are closest together, and slowest in gases, where molecules are farthest apart. **Table 1** shows the speed of sound through different materials.

The Speed of Sound and Temperature The temperature of the material that sound waves are traveling through also affects the speed of sound. As a substance heats up, its molecules move faster, so they collide more frequently. The more frequent the collisions are, the faster the speed of sound is in the material. For example, the speed of sound in air at 0°C is 331 m/s; at 20°C, it is 343 m/s.

Intensity and Loudness

What's the difference between loud sounds and quiet sounds? Imagine a small, square loop next to your ear. If you could measure the amount of energy a sound wave carries through the loop in 1 s, you would be measuring the intensity of the sound wave. The **loudness** of a sound is the human perception of the intensity of the sound waves that strike the ears.

Sound waves spread out as they travel, so the energy carried by the wave is spread over an ever-increasing area. As a result, the intensity of a sound wave decreases as it travels farther from its source. For example, the loudness of a person's voice decreases as you move farther away.

Table 1 Speed of Sound in Different Materials	
Material	**Speed (m/s)**
Air	343
Water	1,482
Glass	5,640
Steel	5,960

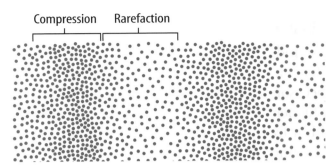

A This sound wave has a lower amplitude.

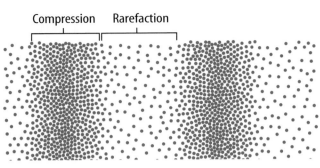

B This sound wave has a higher amplitude. Particles in the material are more compressed in the compressions and more spread out in the rarefactions.

Amplitude and Energy The loudness of a sound depends on how close you are to the sound's source. It also depends on the amount of energy the sound wave carries. The amount of energy a wave carries depends on its amplitude.

For a compressional wave, such as a sound wave, the amplitude is related to how spread out the molecules are in the compressions and rarefactions, as shown in **Figure 4.** The higher the amplitude of the wave is, the more compressed the particles in the compressions are, and the more spread out they are in the rarefactions. More energy had to be transferred by the vibrating object that created the wave to force the particles closer together or spread them farther apart.

Figure 4
The amplitude of a sound wave depends on how spread out the particles are in the wave.

Figure 5
The loudness of sound is measured on the decibel scale.

> ✔ **Reading Check** *What determines the loudness of different sounds?*

The Decibel Scale Perhaps an adult has said to you, "Turn down your music, it's too loud! You're going to lose your hearing!" Although the perception of loudness varies from person to person, the energy carried by sounds waves can be described by a scale called the decibel (dB) scale. **Figure 5** shows the decibel scale. An increase of 10 dB on the decibel scale means the intensity of a sound has increased by 10 times. However, an increase of 20 dB means the intensity has increased by 100 times.

Hearing damage begins to occur at sound levels of about 85 dB. The amount of damage depends on the frequencies of the sound and the length of time a person is exposed to the sound. Some music concerts produce sound levels as high as 120 dB. The intensity of these sound waves is about 30 billion times greater than the intensity of sound waves that are made by whispering.

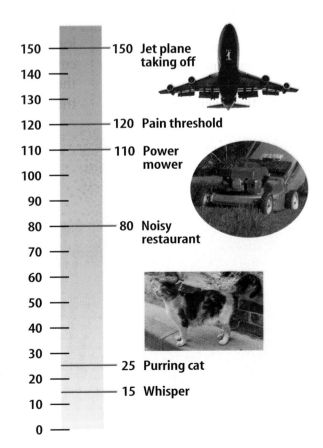

150 — 150 Jet plane taking off
140 —
130 —
120 — 120 Pain threshold
110 — 110 Power mower
100 —
90 —
80 — 80 Noisy restaurant
70 —
60 —
50 —
40 —
30 —
25 Purring cat
20 —
15 Whisper
10 —
0 —

The Doppler Effect

Perhaps you've heard an ambulance siren as the ambulance speeds toward you, then goes past. You might have noticed that the pitch of the siren gets higher as the ambulance moves toward you. Then as the ambulance moves away, the pitch of the siren gets lower. This change in frequency that is due to the motion of a source of sound is called the **Doppler effect. Figure 9** shows why the Doppler effect occurs.

The Doppler effect occurs whether the sound source or the listener is moving. If you drive past a factory as its whistle blows, the whistle will sound higher pitched as you approach. As you move closer you encounter each sound wave a little earlier than you would if you were sitting still, so the whistle has a higher pitch. When you move away from the whistle, each sound wave takes a little longer to reach you. You hear fewer wavelengths per second, which makes the sound lower in pitch.

Radar guns that are used to measure the speed of cars and baseball pitches also use the Doppler effect. Instead of a sound wave, the radar gun sends out a radar wave. When the radar wave is reflected, its frequency changes depending on the speed of the object, and whether it is moving toward the gun or away from it. The radar gun uses the change in frequency of the reflected wave to determine the object's speed.

Problem-Solving Activity

How does Doppler radar work?

Doppler radar is used by the National Weather Service to detect areas of precipitation and to measure the speed at which a storm moves. Because the wind moves the rain, Doppler radar can "see" into a strong storm and expose the winds. Tornadoes that might be forming in the storm then can be identified.

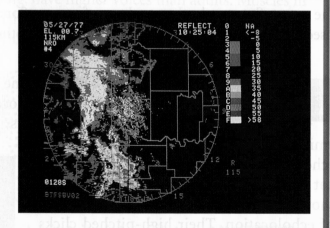

Identify the Problem

An antenna sends out pulses of radio waves as it rotates. The waves bounce off raindrops and return to the antenna at a different frequency, depending on whether the rain is moving toward the antenna or away from it. The change in frequency is due to the Doppler shift.

Solving the Problem

1. If the frequency of the reflected radio waves increases, how is the rain moving relative to the radar station?
2. In a tornado, winds are rotating. How would the radio waves reflected by rotating winds be Doppler-shifted?

Figure 9

You've probably heard the siren of an ambulance as it races through the streets. The sound of the siren seems to be higher in pitch as the ambulance approaches and lower in pitch as it moves away. This is the Doppler effect, which occurs when a listener and a source of sound waves are moving relative to each other.

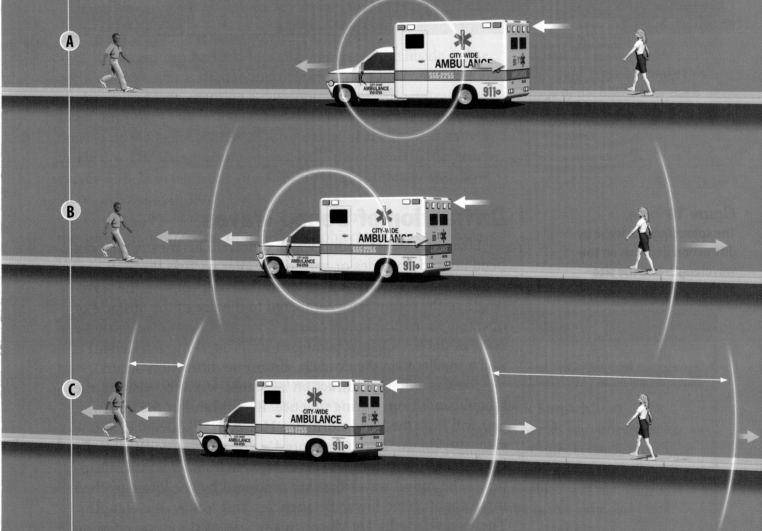

A As the ambulance speeds down the street, its siren emits sound waves. Suppose the siren emits the compression part of a sound wave as it goes past the girl.

B As the ambulance continues moving, it emits another compression. Meanwhile, the first compression spreads out from the point from which it was emitted.

C The waves traveling in the direction that the ambulance is moving have compressions closer together. As a result, the wavelength is shorter and the boy hears a higher frequency sound as the ambulance moves toward him. The waves traveling in the opposite direction have compressions that are farther apart. The wavelength is longer and the girl hears a lower frequency sound as the ambulance moves away from her.

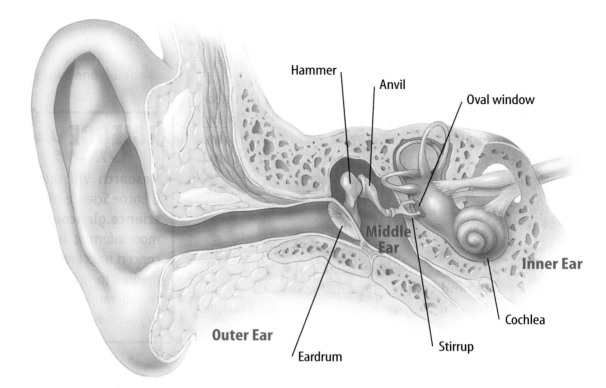

Hammer

Anvil

Oval window

Middle Ear

Inner Ear

Cochlea

Outer Ear

Stirrup

Eardrum

Figure 22
The human ear has three different parts—the outer ear, the middle ear, and the inner ear.

Figure 23
Animals, such as rabbits and owls, have ears that are adapted to their different needs.

The Ear

Sound is all around you. Sounds are as different as the loud buzz of an alarm clock and the quiet hum of a bee. You hear sounds with your ears. The ear is a complex organ that is able to detect a wide range of sounds. The human ear is illustrated in **Figure 22.** It has three parts—the outer ear, the middle ear, and the inner ear.

The Outer Ear—Sound Collector Your outer ear collects sound waves and directs them into the ear canal. Notice that your outer ear is shaped roughly like a funnel. This shape helps collect sound waves.

Animals that rely on hearing to locate predators or prey often have larger, more adjustable ears than humans, as shown in **Figure 23.** A barn owl, which relies on its excellent hearing for hunting at night, does not have outer ears made of flesh. Instead, the arrangement of its facial feathers helps direct sound to its ears. Sea mammals, on the other hand, have small holes for outer ears, even though their hearing is good.

The Middle Ear—Sound Amplifier When sound waves reach the middle ear, they vibrate the **eardrum,** which is a membrane that stretches across the ear canal like a drumhead. When the eardrum vibrates, it passes the vibration to three small connected bones—the hammer, anvil, and stirrup. The bones amplify the vibration of the sound wave, just as a lever can increase a small movement at one end into a larger movement at the other.

The Inner Ear—Sound Interpreter The stirrup vibrates a second membrane called the oval window. This marks the start of the inner ear, which is filled with fluid. Vibrations in the fluid are transmitted to hair-tipped cells lining the cochlea, as shown in **Figure 24.** Different sounds vibrate the cells in different ways. The cells generate signals containing information about the frequency, intensity, and duration of the sound. The nerve impulses travel along the auditory nerve and are transmitted to the part of the brain that is responsible for hearing.

✔ Reading Check *Where are waves detected and interpreted in the ear?*

Hearing Loss

The ear can be damaged by disease, age, and exposure to loud sounds. For example, constant exposure to loud noise can damage hair cells in the cochlea. If these damaged hair cells die, some loss of hearing results because hair cells are not replaced. Hair cells also die as people age. By age 65, most people have lost about 40 percent of these cells and the ability to hear some frequencies is reduced.

The higher frequencies are usually the first to be lost. The loss of the higher frequencies also distorts sound. The soft consonant sounds, such as those made by the letters *s, f, h, sh,* and *ch,* are hard to hear. People with high-frequency hearing loss have trouble distinguishing these sounds in ordinary conversation.

Figure 24
The inner ear contains tiny hair cells that convert vibrations into nerve impulses that travel to the brain.

Section Assessment

1. How are music and noise different?

2. Two bars on a xylophone are 10 cm and 14 cm long. Which bar will produce a lower pitch when struck?

3. Why would the sound of a guitar string sound louder when attached to the body of the guitar than when plucked alone?

4. What are the parts of the human ear, and how do they enable you to hear sound?

5. **Think Critically** As the size of stringed instruments increases from violin to viola, cello, and base, the sound of the instrument becomes lower pitched. Explain.

Skill Builder Activities

6. **Making Models** Illustrate the fundamental and first overtone for a string. **For more help, refer to the** Science Skill Handbook.

7. **Communicating** Imagine that human hearing is much more sensitive than it currently is. Write a story describing a day in the life of your main character. Be sure to describe your setting in detail. For example, does your story take place in a crowded city or a scenic national park? How would life be different? Describe your story in your Science Journal. **For more help, refer to the** Science Skill Handbook.

Music

The pitch of a note that is played on an instrument sometimes depends on the length of the string, the air column, or some other vibrating part. Exactly how does sound correspond to the size or length of the vibrating part? Is this true for different instruments?

Recognize the Problem

What causes different instruments to produce different notes?

Form a Hypothesis

Based on your reading and observations, make a hypothesis about what changes in an instrument to produce different notes.

Goals
- **Design** an experiment to compare the changes that are needed in different instruments to produce a variety of different notes.
- **Observe** which changes are made when playing different notes.
- **Measure and record** these changes whenever possible.

Possible Materials
musical instruments
measuring tape
tuning forks

Safety Precautions
Be sure to wash the mouthpiece of any wind instrument before passing it on to another student.

Test Your Hypothesis

Plan

1. You should do this activity as a class, using as many instruments as possible. You might want to go to the music room or invite friends and relatives who play an instrument to visit the class.

2. As a group, decide how you will measure changes in instruments. Can you measure length in all the instruments you've found? Can you measure thickness or tightness of strings? Two or more should play on each instrument.

3. Refer to the table of wavelengths and frequencies for notes in the scale. Note that no measurements are given—if you measure C to correspond to a string length of 30 cm, for example, the note G will correspond to two thirds of that length.

4. Decide which musical notes you will compare. Prepare a table to collect your data. List the notes you have selected.

Do

1. Make sure your teacher approves your plan before you start.

2. Carry out the experiment as planned.

3. While doing the experiment, record your observations and complete the data table.

Ratios of Wavelengths and Frequencies of Musical Notes		
Note	Wavelength	Frequency
C	1	1
D	8/9	9/8
E	4/5	5/4
F	3/4	4/3
G	2/3	3/2
A	3/5	5/3
B	8/15	15/8
C	1/2	2

Analyze Your Data

1. **Compare** the change in each instrument when the two notes are produced.

2. **Compare and contrast** the changes between instruments.

3. What were the controls in this experiment?

4. What were the variables in this experiment?

5. How did you eliminate bias?

Draw Conclusions

1. Did the results support your hypothesis? Explain.

2. **Describe** how you would modify an instrument to increase the pitch of a note that is played.

3. How does tube length relate to pitch?

Communicating Your Data

Demonstrate to another teacher or to family members how the change in the instrument produces a change in sound.

It's a Wrap!

snap!

pop!

pop!

crackle!

snap!

crackle!

snap!

No matter how fast or slow you open a candy wrapper, it always will make the same noise

You're at the movies, and it's the most exciting part of the film. The audience is silent with their eyes riveted to the screen. At that moment, you decide to unwrap the candy bar you got at the concession stand—"CRACKLE!" The loud noise isn't from the movie. It's from the candy wrapper. Your friends shush you. So you try to open the wrapper more carefully—"POP!" Now you try opening it more slowly—"SNAP!" No matter how you open the candy wrapper— fast or slow—it makes a lot of annoying noise.

Just about everyone has been in that situation at a movie or a concert. And just about everyone has wondered why you can't unwrap candy without making a racket—no matter how hard you try. But now, finally, thanks to the work of a few curious physicists, we know the answer.

To test the plastic problem, researchers took some crinkly wrappers and put them in a silent room. Then the researchers stretched out the wrappers and recorded the sound they made. Next, the crinkling sound was run through a computer. After analyzing the sound, the research team discovered something very interesting—the wrapper didn't make a nonstop, continuous sound. Instead, it made many little separate popping noises. Each of these sound bursts took only a thousandth of a second.

Pop Goes the Wrapper

The researchers found that the loudness of the pops had nothing to do with how fast the plastic was unwrapped. The pops randomly took place. The reason? Little creases in the plastic suddenly snapped into a new position as the wrapper was stretched.

So, if you unwrap candy more slowly, the time between pops will be longer, but the amount of noise made by the pops will be the same. And whether you open the wrapper fast or slow, you'll always hear pops. "And there's nothing you can do about it," said a member of the research team.

Is there another payoff to the candy wrapper research? One scientist said that by understanding what makes a plastic wrapper "snap" when it changes shape, the information can actually help doctors understand molecules in the human body. These molecules, like plastic, can change shape.

But, in the meantime, what are you supposed to do when you absolutely have to open candy in a silent theater? Be considerate of others in the audience. Open the candy as fast as you can, and just get it over with. You can even wait until a noisy part of the movie to hide the crinkle, or open the candy before the film begins.

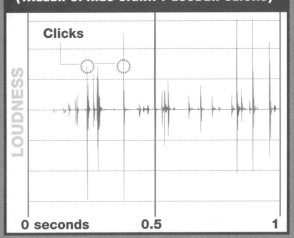

The pop chart

SOUND LEVEL OVER TIME

The sound that a candy wrapper makes is emitted as a series of pulses or clicks. So, opening a wrapper slowly only increases the length of time in between clicks, but the amount of noise remains the same. **(TALLER SPIKES SIGNIFY LOUDER CLICKS)**

Clicks

LOUDNESS

0 seconds   0.5   1

Source: Eric Kramer, Simon's Rock College, 2000

CONNECTIONS Recall and Retell Have you ever opened a candy wrapper in a quiet place? Did it bother other people? If so, did you try to open it more slowly? What happened?

SCIENCE
Online

For more information, visit
science.glencoe.com

Reviewing Main Ideas

Section 1 What is sound?

1. Sound is a compressional wave that travels through matter, such as air. Sound is produced by something that vibrates.

2. The speed of sound is different in different materials. In general, sound travels faster in solids than in liquids, and faster in liquids than in gases. *Will the sound of a train travel faster through the air or through these tracks?*

3. The loudness of a sound depends on the sound's intensity. The intensity increases as the energy carried by the sound wave increases.

4. The pitch of a sound wave corresponds to its frequency. Sound waves can reflect, or bounce, and diffract, or bend around, objects.

5. The Doppler effect occurs when the source of sound and the listener are in motion relative to each other. Sound is shifted up or down in pitch. *What happens to the pitch of the train's horn as it approaches the person?*

Section 2 Music

1. Music is made of sounds that are used in a regular pattern. Noise is made of sounds that are irregular and disorganized.

2. Objects vibrate at their natural frequencies. These depend on the shape of the object and the material it's made of.

3. Resonance occurs when an object is made to vibrate by absorbing energy at one of its natural frequencies.

4. Musical instruments produce notes by vibrating at their natural frequencies. Resonance is used to amplify the sound. *How does resonance make this violin sound louder?*

5 Beats occur when two sounds of nearly the same frequency interfere. The beat frequency is the difference in frequency of the sounds.

6. The ear collects sound waves, amplifies the vibrations, and converts the vibrations to nerve impulses.

FOLDABLES
Reading & Study Skills

After You Read

Use the library to find answers to any questions remaining on your Question Study Foldable.

Visualizing Main Ideas

Complete the following concept map on sound.

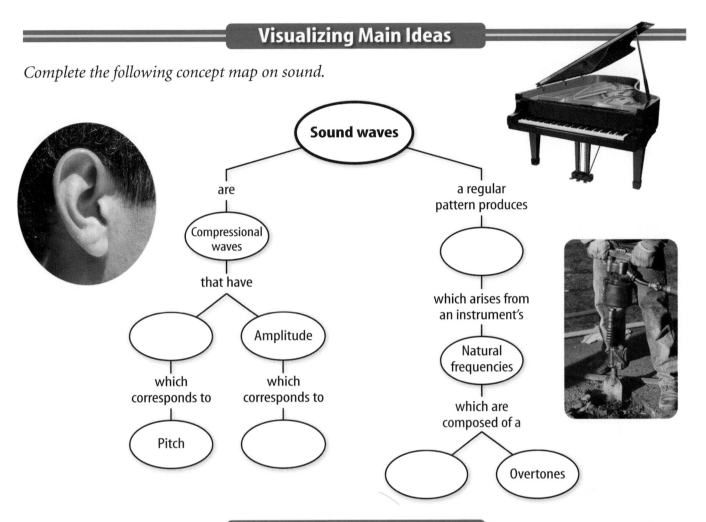

Sound waves

are → Compressional waves

that have → ( ) and Amplitude

which corresponds to → Pitch

which corresponds to → ( )

a regular pattern produces → ( )

which arises from an instrument's → Natural frequencies

which are composed of a → ( ) and Overtones

Vocabulary Review

Vocabulary Words

a. Doppler effect
b. eardrum
c. echo
d. fundamental frequency
e. loudness
f. natural frequency
g. overtone
h. pitch
i. resonance
j. reverberation

THE PRINCETON REVIEW **Study Tip**

Recopy your notes from class. As you do, explain each concept in more detail to make sure that you understand it completely.

Using Vocabulary

Distinguish between the terms in each of the following pairs.

1. overtones, fundamental frequency

2. pitch, sound wave

3. pitch, Doppler effect

4. loudness, resonance

5. fundamental, natural frequency

6. loudness, amplitude

7. natural frequency, overtone

8. reverberation, resonance

Chapter 2 Assessment

Checking Concepts

Choose the word or phrase that best answers the question.

1. A tone that is lower in pitch is lower in what characteristic?
   - **A)** frequency
   - **B)** wavelength
   - **C)** loudness
   - **D)** resonance

2. If frequency increases, what decreases if speed stays the same?
   - **A)** pitch
   - **B)** wavelength
   - **C)** loudness
   - **D)** resonance

3. What part of the ear is damaged by continued exposure to loud noise?
   - **A)** eardrum
   - **B)** stirrup
   - **C)** oval window
   - **D)** hair cells

4. What is an echo?
   - **A)** diffracted sound
   - **B)** resonating sound
   - **C)** reflected sound
   - **D)** Doppler-shifted sound

5. A trumpeter depresses keys to make the column of air resonating in the trumpet shorter. What happens to the note that is being played?
   - **A)** The pitch is higher.
   - **B)** The pitch is lower.
   - **C)** It is quieter.
   - **D)** It is louder.

6. When tuning a violin, a string is tightened. What happens to the note that is being played on that string?
   - **A)** The pitch is higher.
   - **B)** The pitch is lower.
   - **C)** It is quieter.
   - **D)** It is louder.

7. If air becomes warmer, what happens to the speed of sound in air?
   - **A)** It increases.
   - **B)** It decreases.
   - **C)** It doesn't change.
   - **D)** It varies.

8. Sound is what type of wave?
   - **A)** slow
   - **B)** transverse
   - **C)** compressional
   - **D)** fast

9. What does the middle ear do?
   - **A)** focuses sound
   - **B)** interprets sound
   - **C)** collects sound
   - **D)** transmits and amplifies sound

10. An ambulance siren speeds away from you. What happens to the pitch you hear?
    - **A)** It increases.
    - **B)** It becomes louder.
    - **C)** It decreases.
    - **D)** Nothing happens.

Thinking Critically

11. Some xylophones have open pipes of different lengths hung under each bar. The longer the bar is, the longer the corresponding pipe is. Explain how these pipes amplify the sound of the xylophone.

12. Why don't you notice the Doppler effect for a slow-moving train?

13. Suppose the movement of the bones in the middle ear were reduced. Which would be more affected—the ability is hear quiet sounds, or the ability to hear certain frequencies? Explain your answer.

14. Two flutes are playing at the same time. One flute plays a note with frequency 524 Hz. If two beats are heard per second, what are the possible frequencies the other flute is playing?

15. The triangle is a percussion instrument consisting of an open metal triangle hanging from a string. The triangle is struck by a metal rod, and a chiming sound is heard. If the metal triangle is held in the hand rather than by the string, a quiet, dull sound is made when it is struck. Explain why holding the triangle makes it sound quieter.

Developing Skills

16. Predicting If the holes of a flute are all covered while playing, then all uncovered, what happens to the length of the vibrating air column? What happens to the pitch of the note?

17. Identifying and Manipulating Variables and Controls Describe an experiment to demonstrate that sound is diffracted.

18. Making and Using Tables Make a table to show the first three overtones for a note of G (384 Hz).

19. Interpreting Scientific Illustrations The picture shows pan pipes. How are different notes produced on pan pipes?

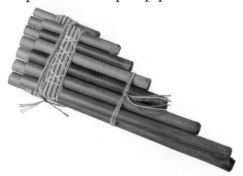

Performance Assessment

20. Recital Perform a short musical piece on an instrument. Explain how your actions changed the notes that were produced.

21. Pamphlet Create a pamphlet describing how a hearing aid works.

TECHNOLOGY

Go to the Glencoe Science Web site at **science.glencoe.com** or use the **Glencoe Science CD-ROM** for additional chapter assessment.

THE PRINCETON REVIEW — Test Practice

Sound travels in waves that change as the pitch and loudness of the sound varies. Here are pictures illustrating four recorded sounds.

Q.

R.

S.

T.

Study the pictures and answer the following questions.

1. Which of the four sounds was getting louder while it was recorded?
   A) Q
   B) R
   C) S
   D) T

2. Which sound had the highest pitch while it was recorded?
   F) Q
   G) R
   H) S
   J) T

Electromagnetic Waves

Wherever you go, you are being bombarded by electromagnetic waves. Some, such as visible light, can be seen. Infrared rays can't be seen but feel warm on your skin. The paint on the tricycle in this picture is being heat cured in an infrared oven. In this chapter, you will learn how electromagnetic waves are formed. You also will learn ways in which electromagnetic waves are used, from cooking to satellite communications.

What do you think?

Science Journal Look at the photograph below with a classmate. Discuss what you think this might be. Here is a hint: *Scientists built this to get a clearer picture.* Write your answer or best guess in your Science Journal.

Light is a type of wave called an electromagnetic wave. You see light every day, but visible light is only one type of electromagnetic wave. Other electromagnetic waves are all around you, but you cannot see them. How can you detect electromagnetic waves that can't be seen with your eyes?

Detecting invisible light

1. Cut a slit 2 cm long and 0.25 cm wide in the center of a sheet of black paper.

2. Cover a window that is in direct sunlight with the paper.

3. Position a glass prism in front of the light coming through the slit so it makes a visible spectrum on the floor or table.

4. Place one thermometer in the spectrum and a second thermometer just beyond the red light.

5. Measure the temperature in each region after 5 min.

Observe

Write a paragraph in your Science Journal comparing the temperatures of the two regions and offer an explanation for the observed temperatures.

Before You Read

FOLDABLES
Reading & Study Skills

Making a Main Ideas Study Fold Make the following Foldable to help you identify the major topics about electromagnetic waves.

1. Stack four sheets of paper in front of you so the short sides are at the top.

2. Slide the top sheet up so that about 4 cm of the next sheet show. Slide each sheet up so about 4 cm of the next sheet show.

3. Fold the sheets top to bottom to form eight tabs. Staple along the topfold.

4. Label the tabs *Electromagnetic Spectrum, Radio Waves, Microwaves, Infrared Rays, Visible Light, Ultraviolet Light, X Rays,* and *Gamma Rays.*

5. As you read the chapter, list the things you learn about these electromagnetic waves under the tabs.

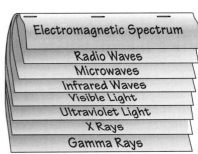

The Nature of Electromagnetic Waves

As You Read

What **You'll Learn**

■ **Explain** how electromagnetic waves are produced.
■ **Describe** the properties of electromagnetic waves.

Vocabulary
electromagnetic wave
radiant energy

Why **It's Important**
Electromagnetic waves provide energy in many useful forms.

Waves in Space

On a clear day you feel the warmth in the Sun's rays, and you see the brightness of its light. Energy is being transferred from the Sun to your skin and eyes. Who would guess that the way in which this energy is transferred has anything to do with radios, televisions, microwave ovens, or the X-ray pictures that are taken by a doctor or dentist? Yet the Sun and the objects shown in **Figure 1** use the same type of wave to move energy from place to place.

Transferring Energy A wave transfers energy from one place to another without transferring matter. How do waves transfer energy? Waves, such as water waves and sound waves, transfer energy by making particles of matter move. The energy is passed along from particle to particle as they collide with their neighbors. Mechanical waves are the types of waves that use matter to transfer energy.

How can a wave transfer energy from the Sun to Earth? Mechanical waves, for example, can't travel in the space between Earth and the Sun where no matter exists. Instead, this energy is carried by a different type of wave called an electromagnetic wave. An **electromagnetic wave** is a wave that can travel through empty space and is produced by charged particles that are in motion.

Figure 1
Getting an X ray at the dentist's office and talking on a cell phone are possible because energy is carried through space by electromagnetic waves.

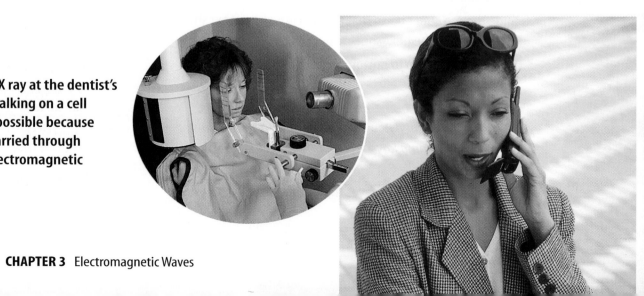

Force due to gravity

Figure 2
A gravity field surrounds all objects, such as Earth.

A When a ball is thrown, Earth's gravity field exerts a downward force on the ball at every point along the ball's path.

B Earth's gravity field extends out through space, exerting a force on all nearby masses.

Forces and Fields

An electromagnetic wave is made of two parts—an electric field and a magnetic field. These fields are force fields. Although you might wonder what a force field is, you already are familiar with one. Earth is surrounded by a force field—gravity. The gravity field surrounding Earth exerts a force on all objects.

> ☑ **Reading Check** *What force field surrounds Earth?*

How does Earth's force field work? If you throw a ball in the air as high as you can, it always falls back to Earth. At every point along the ball's path, the force of gravity pulls down on the ball, as shown in **Figure 2A.** In fact, at every point in space above or at Earth's surface, a ball is acted on by a downward force exerted by Earth's gravity field. The force exerted by this field on a ball could be represented by a downward arrow at any point in space. **Figure 2B** shows this force field that surrounds Earth and extends out into space. In fact, it is Earth's gravity field that causes the Moon to orbit Earth.

Magnetic Fields You know that magnets repel and attract each other even when they aren't touching. Two magnets exert a force on each other when they are some distance apart because each magnet is surrounded by a force field called a magnetic field. Just as a gravity field exerts a force on a mass, a magnetic field exerts a force on another magnet. Magnetic fields cause other magnets to line up along the direction of the magnetic field.

Research In addition to a gravity field, Earth also is surrounded by a magnetic field. Visit the Glencoe Science Web site at **science.glencoe.com** for more information about Earth's gravitational and magnetic force fields. Place the information you gather on a poster to share with your class.

Figure 3
Force fields surround all magnets and charges.

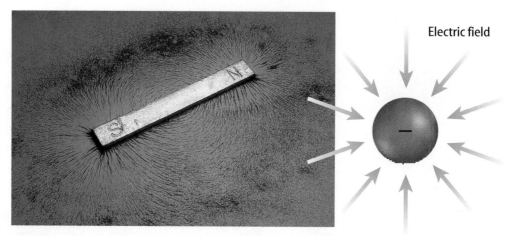

Electric field

A A magnetic field surrounds all magnets. The magnetic field exerts a force on iron filings, causing them to line up with the field.

B The electric field around an electric charge extends out through space, exerting forces on all nearby charged particles.

Electric Fields Recall that atoms contain protons, neutrons, and electrons. Protons and electrons have a property called electric charge. The two types of electric charge are positive and negative. Protons have positive charge and electrons have negative charge.

Just as a magnet is surrounded by a magnetic field, a particle that has electric charge, such as a proton or an electron, is surrounded by an electric field, as shown in **Figure 3.** The electric field is a force field that exerts a force on all other charged particles that are in the field.

Making Electromagnetic Waves

An electromagnetic wave is made of electric and magnetic fields. How is such a wave produced? Think about a wave on a rope. You can make a wave on a rope by shaking one end of the rope up and down. Electromagnetic waves are produced by making charged particles, such as electrons, move back and forth, or vibrate.

A charged particle always is surrounded by an electric field. But a charged particle that is moving also is surrounded by a magnetic field. For example, when an electric current flows in a wire, electrons are moving in the wire. As a result, the wire is surrounded by a magnetic field, as shown in **Figure 4.** So a moving charged particle is surrounded by an electric field and a magnetic field.

Figure 4
Electrons moving in a wire produce a magnetic field in the surrounding space.

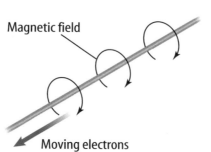

Magnetic field

Moving electrons

Producing Waves When you shake a rope up and down, you produce a wave that moves away from your hand. Likewise, as a charged particle moves up and down, it produces electric and magnetic fields that move away from the vibrating charge. **Figure 5A** shows how these electric and magnetic fields form an electromagnetic wave.

Properties of Electromagnetic Waves

Like all waves, an electromagnetic wave has a frequency and a wavelength. When you create a wave on a rope, you move your hand up and down while holding the rope. Look at **Figure 5B**. Frequency is how many times you move the rope through one complete up and down cycle in 1 s. Wavelength is the distance from one crest to the next or from one trough to the next.

Wavelength and Frequency An electromagnetic wave is produced by a charged particle moving up and down. When the charge makes one complete vibration, one wavelength is created, as shown in **Figure 5A.** Like a wave on a rope, the frequency of an electromagnetic wave is the number of wavelengths that pass by a point in 1 s. This is the same as the number of times in 1 s that the charged particle makes one complete vibration.

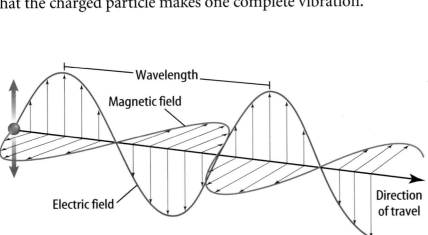

Wavelength

Magnetic field

Electric field

Direction of travel

A When a charged particle moves up, down, and up again, one wavelength of an electromagnetic wave is produced.

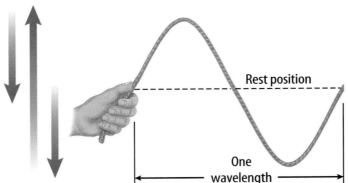

Rest position

One wavelength

Figure 5
Vibrations produce waves, whether they are the result of an electron moving back and forth or a hand shaking a rope up and down.

B By shaking the end of a rope down, up, and down again, you make one wavelength.

Alpha Centauri

Figure 6
The light you see today from Alpha Centauri left the star more than four years ago.

Radiant Energy The energy carried by an electromagnetic wave is called **radiant energy.** What happens if an electromagnetic wave strikes another charged particle? The electric field part of the wave exerts a force on this particle and causes it to move. Some of the radiant energy carried by the wave is transferred into the energy of motion of the particle.

✓ **Reading Check** *What is radiant energy?*

The amount of energy that electromagnetic waves carry is determined by the wave's frequency. The higher the frequency of the electromagnetic wave is, the more energy it has.

The Speed of Light All electromagnetic waves, such as light, microwaves, and X rays, travel through space at the same speed. This speed has been measured as 300,000 km/s in space. Because light is an electromagnetic wave, this speed sometimes is called the speed of light. Electromagnetic waves travel so fast that they could travel around the world more than seven times in 1 s. Even though light travels incredibly fast, it still takes years for light from the stars, other than the sun, to reach Earth. **Figure 6** shows the closest star to the solar system, Alpha Centauri. This star is so far away that the light it emits takes more than four years to reach Earth.

Section 1 Assessment

1. What is an electromagnetic wave?
2. How are electromagnetic waves produced?
3. What two fields surround a moving charged particle?
4. How does the amount of energy carried by a low-frequency wave compare to the amount carried by a high-frequency wave?
5. **Think Critically** Unlike sound waves, electromagnetic waves can travel through a vacuum. What observations can you make to support this statement?

Skill Builder Activities

6. **Comparing and Contrasting** How are electromagnetic waves similar to mechanical waves? How are they different? **For more help, refer to the** Science Skill Handbook.

7. **Calculating Ratios** To go from Earth to Mars, light takes 4 min whereas a spacecraft takes four months. To go to the nearest star, light takes four years. How long would the same spacecraft take to travel to the nearest star? **For more help, refer to the** Math Skill Handbook.

SECTION 2

The Electromagnetic Spectrum

Electromagnetic Waves

The room you are sitting in is bathed in a sea of electromagnetic waves. These electromagnetic waves have a wide range of wavelengths and frequencies. For example, waves with wavelengths from 1 m to 500 m—called radio waves—pass through the walls and windows from distant radio and television broadcast antennas. Other electromagnetic waves—called visible light—have wavelengths more than a million times shorter than radio waves. They are streaming from every object you see.

Classifying Electromagnetic Waves The wide range of electromagnetic waves with different frequencies and wavelengths is called the **electromagnetic spectrum. Figure 7** shows the electromagnetic spectrum. Though many different types of electromagnetic waves exist, they are produced by electric charges that are moving or vibrating. The faster the charge moves or vibrates, the higher the energy of the resulting electromagnetic waves is. Each electromagnetic wave carries radiant energy that increases as the frequency increases. For waves that travel with the same speed, the wavelength increases as frequency decreases. So the energy carried by an electromagnetic wave decreases as the wavelength increases.

As You Read

***What* You'll Learn**
- **Explain** differences among kinds of electromagnetic waves.
- **Identify** uses for different kinds of electromagnetic waves.

Vocabulary

electromagnetic spectrum	ultraviolet radiation
radio wave	X ray
infrared wave	gamma ray
visible light	

***Why* It's Important**
Electromagnetic waves are used to cook food, to send and receive information, and to diagnose medical problems.

Figure 7
Electromagnetic waves have a spectrum of different frequencies.

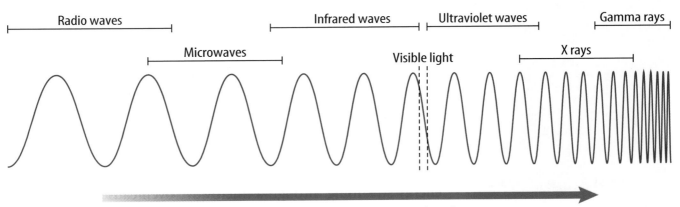

Radio waves   Infrared waves   Ultraviolet waves   Gamma rays

Microwaves   Visible light   X rays

Increasing frequency, decreasing wavelength

Figure 11
A pit viper hunting in the dark can detect the infrared waves that the warm body of its prey emits.

Detecting Infrared Waves Infrared rays are emitted by almost every object. In any material the atoms and molecules are in constant motion. Electrons in the atoms and molecules also move and vibrate. As a result, they give off electromagnetic waves. Most of the electromagnetic waves given off by an object at room temperature are infrared waves and have a wavelength of about 0.000 01 m, or one hundred thousandth of a meter.

Infrared detectors detect objects that are warmer or colder than their environment. They can be used to map wildfires obscured by smoke or to survey underground volcanic activity. Infrared waves also are used to carry signals in some electronics devices, just as radio waves do. Remote controls for TVs and VCRs use infrared rays.

Animals and Infrared Waves Because they are warm, all living things emit infrared radiation. Some animals can detect infrared radiation directly. Piranhas, for example, can see into the infrared, which helps them find prey in the murky Amazon River. Pit vipers, such as rattlesnakes, have special organs just under their eyes that detect changes in infrared waves, as shown in **Figure 11.** If warm prey, such as a mouse, is nearby, the snake can sense and track it, even in the dark.

Visible Light

As the temperature of an object increases, the atoms and molecules move faster. The electrons also move and vibrate faster, and give off electromagnetic waves of higher frequency and shorter wavelength. If the temperature is high enough, the object might glow, as in **Figure 12.** The electromagnetic waves the hot object is emitting are now detectable with your eyes and are called **visible light.** Visible light has wavelengths between about 0.7 and 0.4 millionths of a meter. What you see as different colors are electromagnetic waves of different wavelengths. Red light has the longest wavelength (lowest frequency), and blue light has the shortest wavelength (highest frequency).

Most objects that you see do not give off visible light. They simply reflect the visible light that is emitted by a source of light, such as the Sun or a lightbulb. Some light sources give off visible light because they are at a high temperature.

Figure 12
When objects are heated, their electrons vibrate faster. With enough energy, the vibrating electrons will emit visible light.

Electromagnetic Waves From the Sun

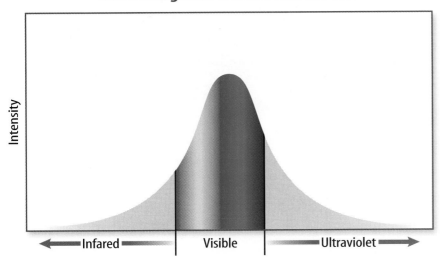

Infared ← | Visible | Ultraviolet →

Figure 13
Electromagnetic waves from the Sun have a range of frequencies centered about the visible region. *Which frequencies of light is the Sun brightest in?*

Ultraviolet Radiation

Ultraviolet radiation is higher in frequency than visible light and has even shorter wavelengths—between 0.4 millionths of a meter and one billionth of a meter. Ultraviolet radiation has higher frequencies than visible light and carries more energy. The radiant energy carried by an ultraviolet wave can be enough to damage the large, fragile molecules that make up living cells. Too much ultraviolet radiation can damage or kill healthy cells.

Figure 13 shows the electromagnetic waves emitted by the Sun, some of which are in the ultraviolet region. Too much exposure to those ultraviolet waves can cause sunburn. Exposure to these waves over a long period of time can lead to early aging of the skin and possibly skin cancer. You can protect yourself from receiving too much ultraviolet radiation by wearing sunglasses and sunscreen, and staying out of the Sun when it is most intense.

Beneficial Uses of UV Radiation
A few minutes of exposure each day to ultraviolet radiation from the Sun enables your body to produce the vitamin D it needs. Most people receive that amount during normal activity. The body's natural defense against too much ultraviolet radiation is to tan. However, a tan can be a sign that overexposure to ultraviolet radiation has occurred.

Ultraviolet radiation's cell-killing effect has led to its use as a disinfectant for surgical equipment in hospitals. In some high school chemistry labs, ultraviolet rays are used to sterilize goggles, as shown in **Figure 14.**

Figure 14
Sterilizing devices, such as this goggle sterilizer, use ultraviolet waves to kill organisms on the equipment.

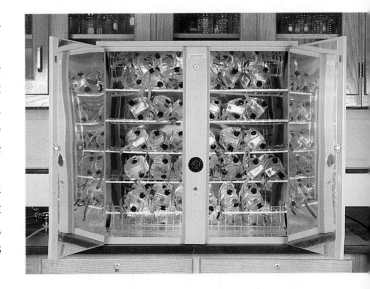

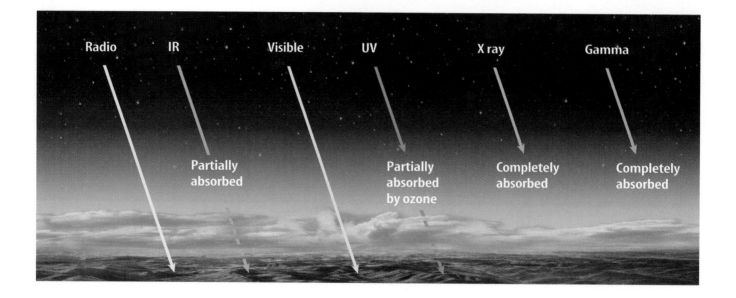

Radio   IR   Visible   UV   X ray   Gamma

Partially absorbed

Partially absorbed by ozone

Completely absorbed

Completely absorbed

Figure 15
Earth's atmosphere serves as a shield to block certain types of electromagnetic waves from reaching the surface.

Life Science
INTEGRATION

Warm-blooded animals, such as mammals, produce their own body heat. Cold-blooded animals, such as reptiles, absorb heat from the environment. Brainstorm the possible advantages of being either-warm-blooded or cold-blooded. Which animals would be easier for a pit viper to detect?

The Ozone Layer Much of the ultraviolet radiation arriving at Earth is absorbed in the upper atmosphere by the ozone layer, as shown in **Figure 15.** Ozone is a molecule that has three oxygen atoms and is formed high in Earth's atmosphere.

However, chemical compounds called CFCs, which are used in air conditioners and refrigerators, can react chemically with ozone. This reaction causes ozone to break down and increases the amount of ultraviolet radiation that penetrates the atmosphere. To prevent this, the use of CFCs is being phased out.

Ultraviolet radiation is not the only type of electromagnetic wave absorbed by Earth's atmosphere. Higher energy waves of X rays and gamma rays also are absorbed. The atmosphere is transparent to radio waves and visible light and partially transparent to infrared waves.

X Rays and Gamma Rays

Ultraviolet rays can penetrate the top layer of your skin. **X rays,** with an even higher frequency than ultraviolet rays, have enough energy to go right through skin and muscle. Heavy lead metal is required to stop the penetrating power of X rays.

Gamma rays have the highest frequency and, therefore, the most penetrating power. They are produced by changes in the nuclei of atoms. When protons and neutrons bond together in nuclear fusion or break apart from each other in nuclear fission, enormous quantities of energy are released. Some of this energy is released as gamma rays.

Just as too much ultraviolet radiation can hurt or kill cells, too much X ray or gamma radiation can have the same effect. Because the energy of the waves is so much higher, the exposure that is needed to cause damage is much less.

Using High-Energy Electromagnetic Radiation The fact that X rays can pass through the human body makes them good for medical diagnosis, as shown in **Figure 16.** X rays pass through the less dense tissues in skin and other organs. These X rays strike a film, creating a shadow image of the denser tissues.

Although the radiation received from getting one medical or dental X ray is not harmful, the cumulative effect of numerous X rays can be dangerous. Doctors therefore try to avoid giving too many X rays. Lead shields or aprons, which protect the internal organs in your body, are used by anyone in the room who is not receiving the X ray. Lead is dense enough to absorb X rays, so they do not pass through the apron into the body. The patient also wears an apron if it will not interfere with the X ray such as when getting a dental X ray.

Using Gamma Rays X rays can harm, but gamma rays can kill. However, gamma rays have beneficial uses, just as X rays do. A beam of gamma rays focused on a cancerous tumor can kill the tumor. Gamma radiation also can cleanse food of disease-causing bacteria. More than 1,500 Americans die each year from *Salmonella* bacteria in poultry and *E. coli* bacteria in meat. Although gamma radiation has been used since 1963 to kill bacteria in food, this method is not widely used in the food insustry.

Astronomy Across the Spectrum

Some astronomical objects produce no visible light and are known only through infrared and radio images. Some galaxies emit X rays from regions that do not emit visible light. Studying astronomy using only visible light would be like looking at only one color in a picture. **Figure 17** shows how different electromagnetic waves can be used to study the universe.

Figure 16
Dense tissues such as bone absorb more X rays than do softer tissues. Consequently, dense tissues leave a shadow on film that can be use to diagnose medical and dental conditions.

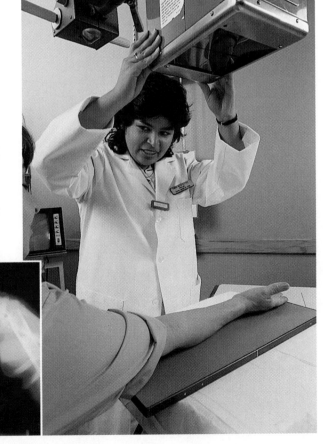

Figure 17

For centuries, astronomers studied the universe using only the visible light coming from planets, moons, and stars. But many objects in space also emit X rays, ultraviolet and infrared radiation, and radio waves. Scientists now use telescopes that can "see" these different types of electromagnetic waves. As these images of the Sun reveal, the new tools are providing remarkable views of objects in the universe.

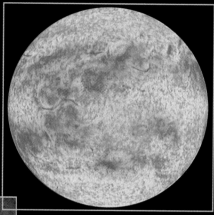

▲ **INFRARED RADIATION** An infrared telescope reveals that the Sun's surface temperature is not uniform. Some areas are hotter than others.

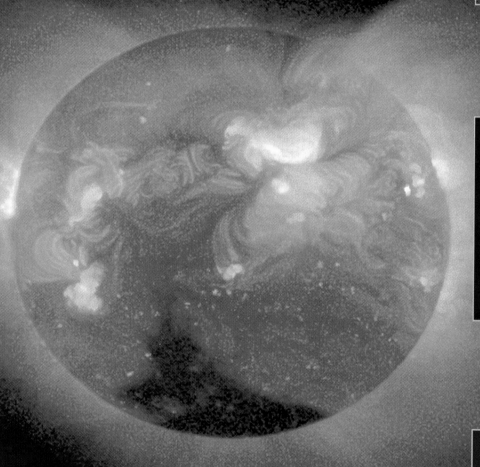

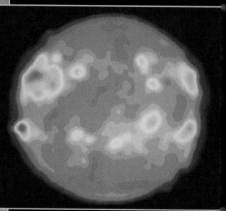

▲ **RADIO WAVES** Radio telescopes detect radio waves given off by the Sun, which have much longer wavelengths than visible light.

▲ **X RAYS** X-ray telescopes can detect the high-energy, short-wavelength X rays produced by the extreme temperatures in the Sun's outer atmosphere.

▶ **ULTRAVIOLET RADIATION** Telescopes sensitive to ultraviolet radiation— electromagnetic waves with shorter wavelengths than visible light—can "see" the Sun's outer atmosphere.

Figure 18
Launching satellite observatories above Earth's atmosphere is the only way to see the universe at electromagnetic wavelengths that are absorbed by Earth's atmosphere.

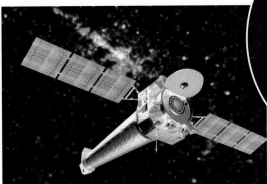

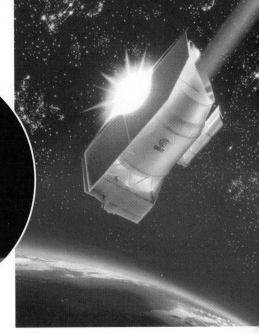

Astronomy INTEGRATION

Satellite Observations Recall from **Figure 15** that Earth's atmosphere blocks many parts of the electromagnetic spectrum. For example, X rays, gamma rays, most ultraviolet rays, and some infrared rays cannot pass through. This is one reason for placing telescopes on satellites. An X-ray image of the galaxy taken from space collects much more information than one taken from Earth's surface. **Figure 18** shows three such satellites—the Extreme Ultraviolet Explorer (EUVE), the Chandra X-Ray Observatory, and the Infrared Space Observatory (ISO).

✔ **Reading Check**   *Why are telescopes sent into space on artificial satellites?*

Section ② Assessment

1. List three types of electromagnetic waves produced by the Sun.

2. Why is ultraviolet light more damaging to cells than infrared light is?

3. Give an application of infrared waves.

4. Describe the difference between X rays and gamma rays.

5. **Think Critically** How is the atmosphere absorbing electromagnetic waves like the process that occurs in a microwave oven?

Skill Builder Activities

6. **Recognizing Cause and Effect** If visible light is the effect, what is the cause? Do the different colors of light have different causes? **For more help, refer to the** Science Skill Handbook.

7. **Using a Database** What do images of the same object look like in different wavelengths? Use a database to research this topic and present a report to your class. **For more help, refer to the** Technology Skill Handbook.

Activity

Prisms of Light

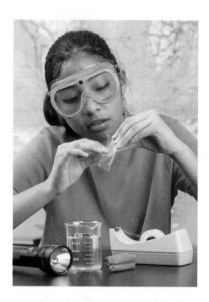

Do you know what light is? Many would answer that light is what you turn on to see at night. However, visible light is made of many different frequencies of the electromagnetic spectrum. A prism can separate visible light into its different frequencies. You see different frequencies of light as different colors. What colors do you see when light passes through a prism?

What You'll Investigate
What happens to visible light as it passes through a prism?

Goals
- **Construct** a prism and observe the different colors that are produced.
- **Infer** how the order of the colors corresponds to the electromagnetic spectrum.

Materials
microscope slides (3)   flashlight
transparent tape   water
clay

Safety Precautions

Procedure
1. Carefully tape the three slides together on their long sides so they form a long prism.
2. Place one end of the prism into a softened piece of clay so the prism is standing upright.
3. Fill the prism with water and put it on a table that is against a dark wall.
4. Shine a flashlight beam through the prism so the light becomes visible on the wall.

Conclude and Apply
1. What was the order of the colors you saw on the wall?
2. **Infer** how a rainbow is created in nature without the use of a prism.
3. What in nature acts similar to a prism to separate visible light?
4. The range of colors, called the spectrum, that you see through the prism is a result of what property of light?
5. The human eye responds best to the green and yellow hues that are found in the middle of the spectrum. Based on the spectrum that is emitted from your prism, infer where infrared light and ultraviolet light are positioned on the electromagnetic spectrum.

*C*ommunicating
Your Data
Compare your conclusions with those of other students in your class. **For more help, refer to the** Science Skill Handbook.

Using Electromagnetic Waves

Telecommunications

In the past week, have you spoken on the phone, watched television, done research on the Internet, or listened to the radio? Today you can talk to someone far away or transmit and receive information over long distances almost instantly. Thanks to telecommunication the world is becoming increasingly connected through the use of electrical impulses and radio waves.

Using Radio Waves

For sending information, the most versatile type of electromagnetic wave to use is radio. Using radio waves to communicate has several advantages. For example, radio waves pass through walls and windows easily. Radio waves do not interact with humans, so they are not harmful to people like ultraviolet rays or X rays are. So for most telecommunication technology, such as TVs, radios, and telephones, radio waves are the electromagnetic wave of choice. **Figure 19** shows the basic method for using radio waves to transmit information—in this case, taking sound from one location and reproducing the sound in a second location.

As You Read

What **You'll Learn**
- **Explain** different methods of electronic communication.
- **Compare and contrast** AM and FM signals.

Vocabulary
Carrier wave
Global Positioning System

Why **It's Important**
Telecommunication enables people to contact others and collect information worldwide.

Figure 19
Radio transmission relies on conversions among sound, electrical, and radiant energies.

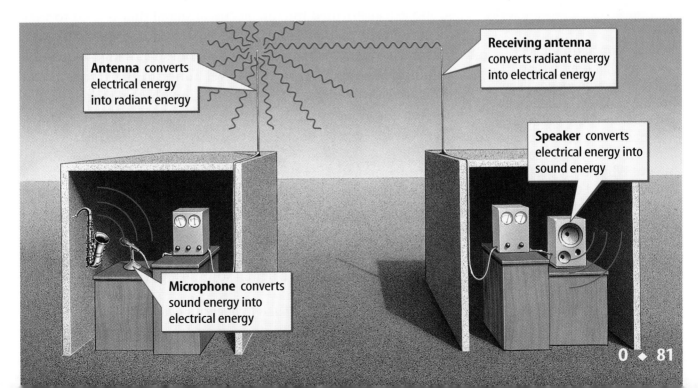

Antenna converts electrical energy into radiant energy

Receiving antenna converts radiant energy into electrical energy

Speaker converts electrical energy into sound energy

Microphone converts sound energy into electrical energy

Figure 20
An information signal can be stored on a carrier wave in two ways—amplitude modulation or frequency modulation.

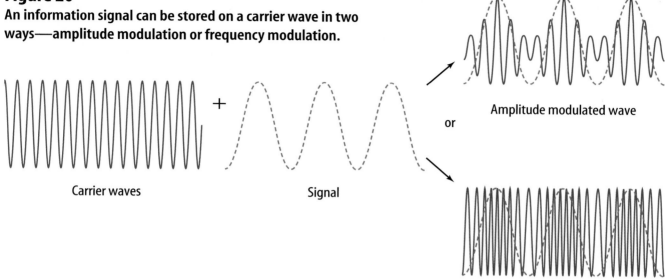

Carrier waves   Signal

or

Amplitude modulated wave

Frequency modulated wave

Radio Transmission The simplest way to transmit a signal is to start and stop it. Morse code uses such an on/off signal. Modern telecommunications use altered waves. Each radio station is assigned one particular frequency at which to transmit, such as 105.1 MHz. The assigned frequency is the **carrier wave** for that station. A carrier wave's frequency is where you tune your radio to receive a particular station's broadcast. To carry information on the carrier wave, either the amplitude or the frequency of the carrier wave is changed, or modulated.

Amplitude Modulation The letters *AM* in AM radio stands for amplitude modulation, which means that the amplitude of the carrier wave is changed to transmit information. The original sound is transformed into an electrical signal that is used to vary the amplitude of the carrier wave, as shown in **Figure 20.** Note that the frequency of the carrier wave doesn't change—only the amplitude changes. An AM receiver tunes to the frequency of the carrier wave. Then the information that is stored in the varying amplitude of the carrier wave creates an electrical signal that goes to the speaker, reproducing the original sound.

Frequency Modulation FM radio works in much the same way as AM radio, but the frequency instead of the amplitude is modulated, as shown in **Figure 20.** In this case, the receiver uses the information stored in the varying frequency of the carrier wave to create an electrical signal. When the electrical signal reaches the speakers, the original sound is reproduced.

 **Reading Check** *What is frequency modulation?*

Astronomy
INTEGRATION

Pulsars are astronomical objects that emit periodic bursts of radio waves. The pattern of pulses is regular. Investigate how pulsars originate and communicate to your class what you learn. Why might pulsars have seemed to be signals from intelligent life?

Telephones

A telephone contains a microphone in the mouthpiece that converts a sound wave into an electric signal. The electric signal is carried through a wire to the telephone network. There, the signal might remain electric or be converted into a radio or microwave signal for transmission through the air. The electric signal also can be converted into a light wave for transmission through fiber-optic cables.

At the receiving end, the signal is converted back to an electric signal. A speaker in the earpiece of the phone changes the electric signal into a sound wave.

☑ Reading Check *What device converts sound into an electric signal?*

Math Skills Activity

Calculating the Wavelength of Radio Frequencies

Example Problem

You are listening to an FM station with a frequency of 94.9 MHz or 94,900,000 Hz. How long are the wavelengths that strike the antenna? For any wave, the wavelength equals the wave speed divided by the frequency. The speed of radio waves is 300,000,000 m/s. The SI unit of frequency, Hz, is equal to l/s.

Solution

1 *This is what you know:*   frequency = 94,900,000 Hz
     wave speed = 300,000,000 ms

2 *This is what you need to find:*   wavelength

3 *This is the equation you need to use:*   wavelength = wave speed / frequency

4 *Substitute the known values:*   wavelength = 300,000,000 m/s / 94,900,000 Hz
             = 3.16 m

Check your answer by multiplying the units. Do you calculate a unit of distance for your answer?

Practice Problems

1. Your friend prefers an AM radio station at 1,520 kHz (1,520 thousand vibrations each second). What is the wavelength of this frequency? Which has a longer wavelength, AM or FM radio waves?

2. An AM radio station operates at 580 kHz (580 thousand vibrations each second). What is the wavelength of this frequency? What is the relationship between frequency and wavelength?

For more help, refer to the Math Skill Handbook.

Figure 21
Electromagnetic waves make using telephones easier.

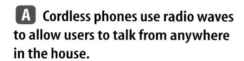 **A** Cordless phones use radio waves to allow users to talk from anywhere in the house.

B Radio waves enable cell phone users to send or receive calls without using wires.

Remote Phones A telephone does not have to transmit its signal through wires. In a cordless phone, the electrical signal produced by the microphone is transmitted through an antenna in the handset to the base. **Figure 21A** shows how incoming signals are transmitted from the base to the handset. A cellular phone uses an antenna to broadcast and receive information between the phone and a base station, as shown in **Figure 21B.** The base station uses radio waves to communicate with other stations in a network.

Pagers The base station also is used in a pager system. When you dial a pager, the signal is sent to a base station. From there, an electromagnetic signal is sent to the pager. The pager beeps or vibrates to indicate that someone has called. With a touch-tone phone, you can transmit numeric information, such as your phone number, which the pager will receive and display.

Communications Satellites

How do you send information to the other side of the world? Radio waves can't be sent directly through Earth. Instead, radio signals are sent to satellites. The satellites can communicate with other satellites or with ground stations. Some communications satellites are in geosynchronous orbit, meaning each satellite remains above the same point on the ground.

The Global Positioning System

Satellites also are used as part of the **Global Positioning System,** or GPS. GPS is used to locate objects on Earth. The system consists of satellites, ground-based stations, and portable units with receivers, as illustrated in **Figure 22.**

GPS measures the time it takes for a user's portable unit to receive radio waves from several satellites. The time shows how far the receiver is from each satellite. By communicating with several satellites, a user's longitude, latitude, and elevation can be determined. Different receivers have different levels of accuracy, giving location to within a few hundred meters for sailors or within centimeters for surveyors working on topographic maps.

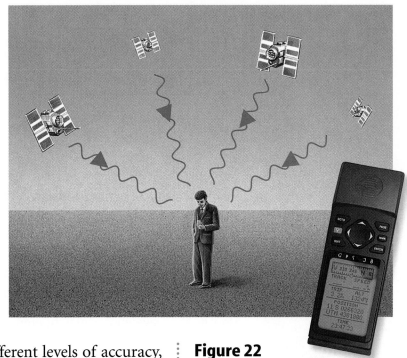

Figure 22
The Global Positioning System (GPS) works by using a series of satellites, ground-based stations, and portable units with receivers.

 **Reading Check** *What is GPS used for?*

Many of these forms of communication have been developed over the past few decades. For example, an Internet connection transfers images and sound using the telephone network, just as a television signal transfers images and sound using radio waves. What forms of telecommunications do you think you'll be using a few decades from now?

Section 3 Assessment

1. What is a modulated radio signal?

2. What does a microphone do? What does a speaker do?

3. What types of information does a GPS receiver provide for its user?

4. What is a communications satellite?

5. **Think Critically** Electromagnetic waves travel more slowly in materials such as glass than through air. What might be the benefit of sending information in glass wires be?

Skill Builder Activities

6. **Researching Information** Find out more about a form of telecommunications, such as email or shortwave radio. **For more help, refer to the** Science Skill Handbook.

7. **Communicating** Think of a story you have enjoyed about a time before telecommunications or one in which telecommunication was not possible. How would telecommunications have changed the story? **For more help, refer to the** Science Skill Handbook.

Spectrum Inspection

You've heard the term "red-hot" used to describe something that is unusually hot. When a piece of metal is heated it may give off a red glow or even a yellow glow. All objects emit electromagnetic waves. How do the wavelengths of these waves depend on the temperature of the object?

Recognize the Problem

How do the wavelengths of light produced by a lightbulb depend on the temperature of the lightbulb?

Form a Hypothesis

The brightness of a lightbulb increases as its temperature increases. Form a hypothesis describing how the wavelengths emitted by a lightbulb will change as the brightness of a lightbulb changes.

Goals
- **Design** an experiment that determines the relationship between brightness and the wavelengths emitted by a lightbulb.
- **Observe** the wavelengths of light emitted by a lightbulb as its brightness changes.

Safety Precautions

WARNING: Be sure all electrical cords and connections are intact and that you have a dry working area. Do not touch the bulbs as they may be hot.

Possible Materials
diffraction grating
power supply with variable resister switch
clear, tubular lightbulb and socket
red, yellow, and blue colored pencils

Test Your Hypothesis

Plan

1. **Decide** how you will determine the effect of lightbulb brightness on the colors of light that are emitted.

2. As shown in the photo at the right, you will look toward the light through the diffraction grating to detect the colors of light emitted by the bulb. The color spectrum will appear to the right and to the left of the bulb.

3. **List** the specific steps you will need to take to test your hypothesis. Describe precisely what you will do in each step. Will you first test the bulb at a bright or dim setting? How many settings will you test? (Try at least three.) How will you record your observations in an organized way?

4. **List** the materials you will need for your experiment. Describe exactly how and in which order you will use these materials.

5. **Identify** any constants and variables in your experiment.

Do

1. Make sure your teacher approves your plan before you start.

2. **Perform** your experiment as planned.

3. While doing your experiment, write down any observations you make in your Science Journal.

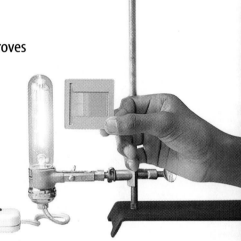

Analyze Your Data

1. Use the colored pencils to draw the color spectrum emitted by each brightness.

2. Which colors appeared as the bulb became brighter? Did any colors disappear?

3. How did the wavelengths emitted by the bulb change as the bulb became brighter?

4. Infer how the wavelengths emitted by the lightbulb changed as it became hotter.

Draw Conclusions

1. If an object becomes hotter, what happens to the wavelengths it emits?

2. How do the wavelengths that the bulb emits change if it is turned off?

3. From your results, infer whether red stars or yellow stars are hotter.

*C*ommunicating Your Data

Compare your results with others in your class. How many different colors were seen?

Hedy Lamarr, actor and inventor

Hopping the

Ringgggg. There it is—that familiar beep! Out come the cellular phones—from purses, pockets, book bags, belt clips, and briefcases. At any given moment, a million wireless signals are flying through the air—and not just cell phone signals. With radio and television signals, Internet data, and even Global Positioning System information coming at us, the air seems like a pretty crowded place. How do all of these signals get to the right place? How does a cellular phone pick out its own signal from among the clutter? The answer lies in a concept developed in 1940 by Hedy Lamarr.

Lamarr was born in Vienna, Austria. In 1937, she left Austria to escape Hitler's invading Nazi army. Lamarr left for another reason, as well. She was determined to pursue a career as an actor. And she became a famous movie star.

In 1940, Lamarr came up with an idea to keep radio signals that guided torpedoes from being jammed. Her idea, called frequency hopping, involved breaking the radio signal that was guiding the torpedo into tiny parts and rapidly changing their frequency. The enemy would not be able to keep up with the changes and thus would not be able to divert the torpedo from its target. Lamarr worked with a partner who helped her figure out how to make the idea work. They were awarded a patent for their idea in 1942.

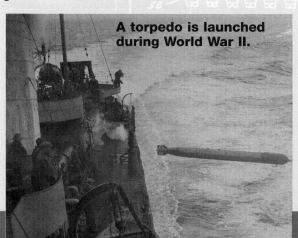

A torpedo is launched during World War II.

Spread Spectrum

Lamarr's idea was ahead of its time. The digital technology that allowed efficient operation of her system wasn't invented until decades later. However, after 1962, frequency hopping was adopted and used in U.S. military communications. It was the development of cellular phones, however, that benefited the most from Lamarr's concept.

Cellular phones and other wireless technologies operate by breaking their signals into smaller parts, called packets. The packets are encoded in a certain way for particular receivers and are spread across bands of the electromagnetic spectrum. In this way, millions of users can use the same frequencies at the same time.

Frequencies

CONNECTIONS Brainstorm **How are you using wireless technology in your life right now? List ways it makes your life easier. Are there drawbacks to some of the uses for wireless technology? What are they?**

SCIENCE
Online

For more information, visit
science.glencoe.com

Reviewing Main Ideas

Section 1 The Nature of Electromagnetic Waves

1. Moving charges generate vibrating electric and magnetic fields. These vibrating fields travel through space and are called electromagnetic waves.

2. Electromagnetic waves, like all waves, have wavelenth, frequency, and energy. *How are ocean waves similar to electromagnetic waves?*

Section 2 The Electromagnetic Spectrum

1. Radio waves have the longest wavelength and lowest energy. Microwaves and radar are subsets of radio waves.

2. All objects emit infrared waves. If you see an object, visible light must be coming from it. *At night, how can a person be detected by an infrared camera?*

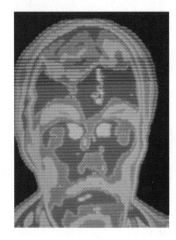

3. Ultraviolet waves have a higher frequency and more energy than visible light.

4. X rays and gamma rays are highly penetrating and can be dangerous to living organisms.

Section 3 Using Electromagnetic Waves

1. Communications systems use visible light, radio waves, or electrical signals to transmit information.

2. Radios use modulated carrier waves to transmit information.

3. Electromagnetic waves are used in telephone technologies to make communication easier and faster. *What is one way an electromagnetic wave is used in telephone communication?*

4. Communications satellites relay information from different points on Earth so a transmission can go around the globe. The Global Positioning System is one application of satellites.

FOLDABLES
Reading & Study Skills

After You Read

Using the information on your Foldable, compare and contrast visible and invisible waves that form the electromagnetic spectrum.

Visualizing Main Ideas

Complete the following spider map about electromagnetic waves.

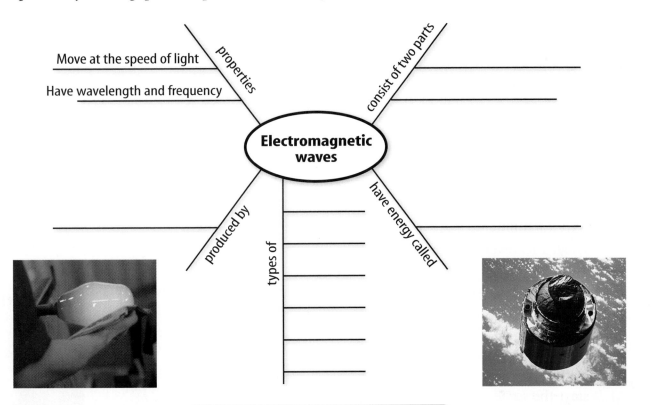

properties
- Move at the speed of light
- Have wavelength and frequency

consist of two parts
_____
_____

Electromagnetic waves

have energy called
_____

types of
_____
_____
_____
_____
_____

produced by
_____

Vocabulary Review

Vocabulary Words

a. carrier wave
b. electromagnetic spectrum
c. electromagnetic wave
d. gamma ray
e. Global Positioning System
f. infrared wave
g. radiant energy
h. radio wave
i. ultraviolet radiation
j. visible light
k. X ray

Using Vocabulary

Explain the difference between the terms in each of the following pairs.

1. infrared wave, radio wave
2. radio wave, carrier wave
3. communications satellite, GPS
4. visible light, ultraviolet radiation
5. X ray, gamma ray
6. electromagnetic spectrum, rainbow
7. X ray, photograph
8. television wave, AM radio wave
9. electromagnetic wave, light
10. infrared wave, ultraviolet wave

THE PRINCETON REVIEW **Study Tip**

After you read a chapter, write ten questions that it answers. Wait one day and then try to recall the answers. Look up what you can't remember.

Checking Concepts

Choose the word or phrase that best answers the question.

1. Which type of force field surrounds a moving electron?
   A) gravity
   C) magnetic
   B) electric
   D) all of these

2. What does a microphone transform?
   A) light waves to sound waves
   B) radio waves to an electrical signal
   C) sound waves to electromagnetic waves
   D) sound waves to an electrical signal

3. Which of the following electromagnetic waves have the lowest frequency?
   A) visible light
   C) radio waves
   B) infrared waves
   D) X rays

4. What happens to the energy of an electromagnetic wave as its frequency increases?
   A) It increases.
   B) It decreases.
   C) It stays the same.
   D) It oscillates up and down.

5. What type of wave do all hot objects emit?
   A) radio
   C) visible
   B) infrared
   D) ultraviolet

6. What can detect radio waves?
   A) film
   C) eyes
   B) antenna
   D) skin

7. What type of wave passes through people?
   A) infrared
   C) ultraviolet
   B) visible
   D) gamma

8. Which color has the lowest frequency?
   A) green
   C) yellow
   B) violet
   D) red

9. What is the key device that allows remote phones to function?
   A) X ray
   C) GPS
   B) satellite
   D) antenna

10. What does *A* in AM stand for?
    A) amplitude
    C) astronomical
    B) antenna
    D) Alpha centauri

Thinking Critically

11. Infrared light was discovered when a scientist placed a thermometer in each band of the light spectrum produced by a prism. Would the area just beyond red have been warmer or cooler than the room? Explain.

12. Astronomers have built telescopes on Earth that have flexible mirrors that can eliminate the distortions due to the atmosphere. What advantages would a space-based telescope have over these?

13. Heated objects often give off visible light of a particular color. Explain why an object that glows bluish-white is hotter than one that glows red.

14. How can an X ray be used to determine the location of a cancerous tumor?

15. Why are many communications systems based on radio waves?

Developing Skills

16. **Calculating Ratios** How far does light travel in 1 min? How does this compare with the distance to the Moon?

17. **Recognizing Cause and Effect** As you ride in the car, the radio alternates between two different stations. How can the antenna pick up two stations at once?

18. **Classifying** List the colors of the visible spectrum in order of increasing frequency.

19. Comparing and Contrasting Compare and contrast ultraviolet and infrared light.

20. Concept Mapping Electromagnetic waves are grouped according to their frequencies. In the following concept map, write each frequency group and one way humans make use of the electromagnetic waves in that group. For example, in the second set of ovals, you might write "X rays" and "to see inside humans."

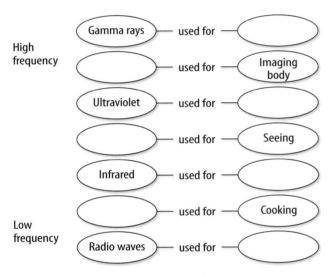

Mr. Rubama's class was studying how radio waves are transmitted. An experimental setup involving radio waves and glass is shown below.

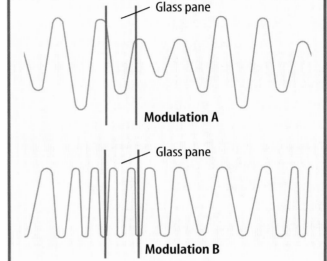

Study the illustrations and answer the following questions.

1. Which of these questions would most likely be answered by this experiment?
   A) How fast do radio waves travel through the air?
   B) Why do some waves travel more quickly than other waves?
   C) Where do radio waves come from?
   D) Can radio waves travel through glass?

2. Which of the following describes how the wave in Modulation A is different from the wave in Modulation B?
   F) It is a radio wave.
   G) It is frequency modulated.
   H) It is amplitude modulated.
   J) It is a a carrier wave.

Performance Assessment

21. Oral Presentation Explain to the class how a radio signal is generated, transmitted, and received.

22. Poster Make a poster showing the parts of the electromagnetic spectrum. Show how frequency, wavelength, and energy change throughout the spectrum. How is each wave generated? What are some uses of each?

Technology

Go to the Glencoe Science Web site at **science.glencoe.com** or use the **Glencoe Science CD-ROM** for additional chapter assessment.

Light, Mirrors, and Lenses

You walk through a door of the fun house and are bombarded by images of yourself. In one mirror, your face seems smashed. You turn around and face another mirror—your chin and neck are gigantic. How do mirrors in a fun house make you look so strange? In this chapter, you'll learn how mirrors and lenses create images. You'll also learn why objects have the colors they have, and how you see.

What do you think?

Science Journal Look at the picture below with a classmate. Discuss what you think this might be or what is happening. Here's a hint: *It helps you keep in touch.* Write down your answer or your best guess in your Science Journal.

Everything you see results from light waves entering your eyes. These light waves are either given off by objects, such as the Sun and lightbulbs, or reflected by objects, such as trees, books, and people. Lenses and mirrors can cause light to change direction and make objects seem larger or smaller. What happens to light as it passes from one material to another?

Observe the bending of light

1. Place two paper cups next to each other and put a penny in the bottom of each cup.

2. Fill one of the cups with water and observe how the penny looks.

3. Looking straight down at the cups, slide the cup with no water away from you just until you can no longer see the penny.

4. Pour water into this cup and observe what seems to happen to the penny.

Observe

In your Science Journal, record your observations. Did adding water make the cup look deeper or shallower?

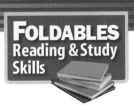

Before You Read

FOLDABLES
Reading & Study Skills

Making a Question Study Fold Asking yourself questions helps you stay focused so you will better understand light, mirrors, and lenses when you are reading the chapter.

1. Stack two sheets of paper in front of you so the short side of both sheets is at the top.

2. Slide the top sheet up so about 4 cm of the bottom sheet show.

3. Fold both sheets top to bottom to form four tabs and staple along the fold as shown.

4. Title the Foldable *Light, Mirrors, and Lenses* as shown. Write these questions on the flaps: *What are the properties of light? What is reflection? What is refraction?*

5. Before you read the chapter, try to answer the questions with what you already know. As you read the chapter, add to or correct your answers under the flaps.

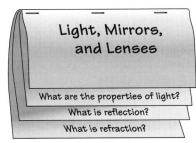

Properties of Light

What You'll Learn

- **Describe** the wave nature of light.
- **Explain** how light interacts with materials.
- **Determine** why objects appear to have color.

Vocabulary

light ray
medium
reflection

Why It's Important

Most of what you know about your surroundings comes from information carried by light waves.

Figure 1
Waves spread in all directions and carry energy.

What is light?

Drop a rock on the smooth surface of a pond, and you'll see ripples spread outward from the spot where the rock struck. The rock produced a wave much like the one in **Figure 1A.** A wave is a rhythmic disturbance that carries energy through matter or space. The matter in this case is the water, and the energy comes from the impact of the rock. As the ripples spread out, they carry some of that energy.

Light is a type of wave that carries energy. A source of light such as the Sun or a lightbulb gives off light waves into space, just as the rock hitting the pond causes waves to form in the water. But while the water waves spread out only on the surface of the pond, light waves spread out in all directions from the light source. **Figure 1B** shows how light waves travel.

Sometimes, however, it is easier to think of light in a different way. A **light ray** is a narrow beam of light that travels in a straight line. You can think of a source of light as giving off, or emitting, a countless number of light rays that are traveling away from the source in all directions.

A Ripples on the surface of a pond are produced by an object hitting the water. As the ripples spread out from the point of impact, they carry energy.

B A source of light, such as a lightbulb, gives off light rays that travel away from the light source in all directions.

Light Travels Through Space There is, however, one important difference between light waves and the water wave ripples on a pond. If the pond dried up and had no water, ripples could not form. Waves on a pond need a material—water—in which to travel. The material through which a wave travels is called a **medium.** Light is an electromagnetic wave and doesn't need a medium in which to travel. Electromagnetic waves can travel in a vacuum, as well as through materials such as air, water, and glass.

Light and Matter

What can you see when you are in a closed room with no windows and the lights are out? You can see nothing until you turn on a light or open a door to let in light from outside the room. Most objects do not give off light on their own. They can be seen only if light waves from another source bounce off them and into your eyes, as shown in **Figure 2.** The process of light striking an object and bouncing off is called **reflection.** Right now, you can see these words because light emitted by a source of light is reflecting from the page and into your eyes. Not all the light rays reflected from the page strike your eyes. Light rays striking the page are reflected in many directions, and only some of these rays enter your eyes.

✓ Reading Check *What must happen for you to see most objects?*

TRY AT HOME

Mini LAB

Observing Colors in the Dark

Procedure
1. Get six pieces of **paper** that are different colors and about 10 cm × 10 cm.
2. Darken a room and wait 10 min for your eyes to adjust to the darkness.
3. Write on each paper what color you think the paper is.
4. Turn on the lights and see if your night vision detected the colors.

Analysis
1. If the room were perfectly dark, what would you see? Explain.
2. Your eyes contain structures called rods and cones. Rods don't detect color, but need only a little light. Cones detect color, but need more light. Which structure was working in the dark room? Explain.

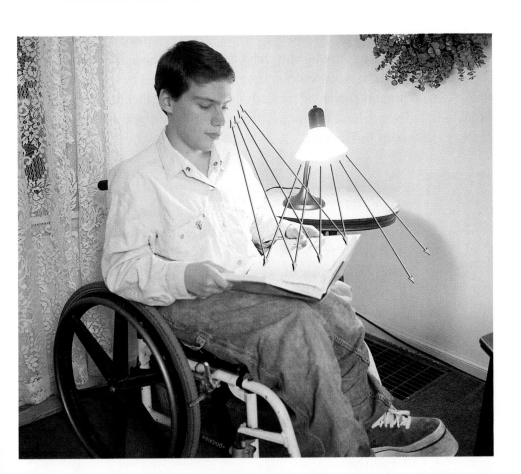

Figure 2
Light waves are given off by the lightbulb. Some of these light waves hit the page and are reflected. The student sees the page when some of these reflected waves enter the student's eyes.

A An opaque object allows no light to pass through it.

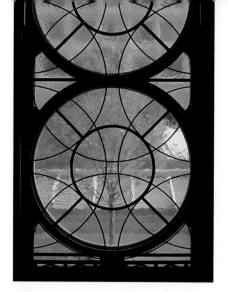

B A translucent object allows some light to pass through it.

C A transparent object allows almost all light to pass through it.

Figure 3
Materials are opaque, translucent, or transparent depending on how much light passes through them. *Which type of material reflects the least amount of light?*

Figure 4
A beam of white light passing through a prism is separated into many colors. *What colors can you see emerging from the prism?*

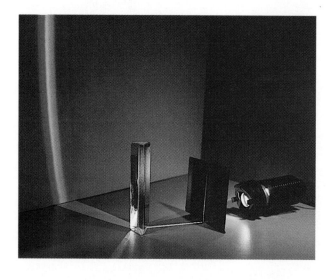

Opaque, Translucent, and Transparent When light waves strike an object, some of the waves are absorbed by it, some of the waves are reflected by it, and some of the light waves might pass straight through it. What happens to light when it strikes the object depends on the material that the object is made of.

All objects reflect and absorb some light waves. Materials that let no light pass through them are opaque (oh PAYK). You cannot see other objects through opaque materials. On the other hand, you clearly can see other objects through materials such as glass and clear plastic that allow nearly all the light that strikes them to pass through. These materials are transparent. A third type of material allows only some light to pass through. Although objects behind these materials are visible, they are not clear. These materials, such as waxed paper or frosted glass, are translucent (trans LEW sent). Examples of opaque, translucent, and transparent objects are shown in **Figure 3.**

Color

The light from the Sun might look white, but it is a mixture of colors. Each different color of light is a different wavelength. You sometimes can see the different colors of the Sun's light when it passes through raindrops to make a rainbow. As shown in **Figure 4,** white light is separated into different colors when it passes through a prism. The colors in white light range from red to violet. When light waves from all these colors enter the eye at the same time, the brain interprets the mixture as being white.

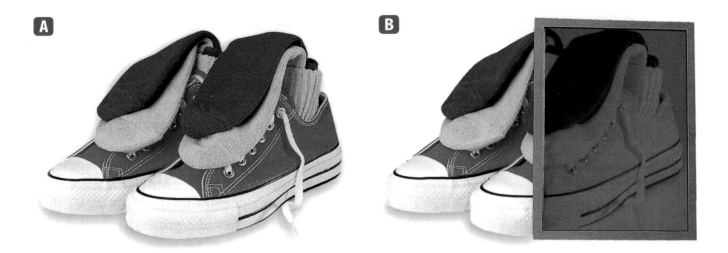

A

B

Why do Objects Have Color?

Why does grass look green or a rose look red? When a mixture of light waves strikes an object that is not transparent, the object absorbs some of the light waves. Some of the light waves that are not absorbed are reflected. If an object reflects red waves and absorbs all the other waves, it looks red. Similarly, if an object looks blue, it reflects only blue light waves and absorbs all the others. An object that reflects all the light waves that strike it looks white, while one that reflects none of the light waves that strike it looks black. **Figure 5** shows gym shoes and socks as seen under white light and as seen when viewed through a red filter that allows only red light to pass through it.

Primary Light Colors

How many colors exist? People often say white light is made up of red, orange, yellow, green, blue, and violet light. This isn't completely true, though. Many more colors than this exist. In reality, humans can distinguish thousands of colors, including some such as brown, pink, and purple, that are not found among the colors of the rainbow.

Light of almost any color can be made by mixing different amounts of red, green, and blue light. Red, green, and blue are known as the primary colors. Look at **Figure 6.** White light is produced where beams of red, green, and blue light overlap. Yellow light is produced where red and green light overlap. You see the color yellow because of the way your brain interprets the combination of the red and green light striking your eye. This combination of light waves looks the same as yellow light produced by a prism, even though these light waves have only a single wavelength.

Figure 5
A Examine the pair of gym shoes and socks as they are seen under white light. *Why do the socks look blue under white light?*
B The same shoes and socks were photographed through a red filter. *Why do the blue socks look black when viewed under red light?*

Figure 6
By mixing light from the three primary colors—red, blue, and green—all of the visible colors can be made.

Figure 9
A highly magnified view of the surface of a paper towel shows that the surface is made of many cellulose wood fibers that make it rough and uneven.

Magnification: 35×

Regular and Diffuse Reflection Even though the surface of the paper might seem smooth, it's not as smooth as the surface of a mirror. **Figure 9** shows how rough the surface of a piece of paper looks when it is viewed under a microscope. The rough surface causes light rays to be reflected from it in many directions, as shown in **Figure 10A.** This uneven reflection of light waves from a rough surface is diffuse reflection. The smoother surfaces of mirrors, as shown in **Figure 10B,** reflect light waves in a much more regular way. For example, parallel rays remain parallel after they are reflected from a mirror. Reflection from mirrors is known as regular reflection. Light waves that are regularly reflected from a surface form the image you see in a mirror or any other smooth surface. Whether a surface is smooth or rough, every light ray that strikes it obeys the law of reflection.

✔ **Reading Check** *Why does a rough surface cause a diffuse reflection?*

Scattering of Light When diffuse reflection occurs, light waves that were traveling in a single direction are reflected, and then travel in many different directions. Scattering occurs when light waves traveling in one direction are made to travel in many different directions. Scattering also can occur when light waves strike small particles, such as dust. You may have seen dust particles floating in a beam of sunlight. When the light waves in the sunbeam strike a dust particle, they are scattered in all directions. You see the dust particle as bright specks of light when some of these scattered light waves enter your eye.

Figure 10
A A rough surface causes parallel light rays to be reflected in many different directions.
B A smooth surface causes parallel light rays to be reflected in a single direction.

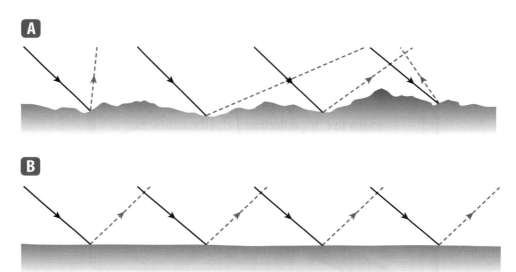

A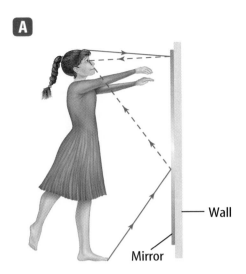

Wall

Mirror

B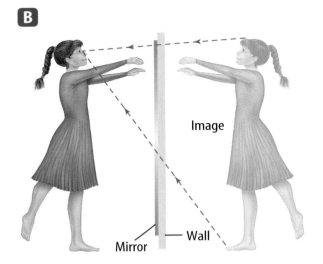

Image

Mirror — Wall

Reflection by Plane Mirrors Did you glance in the mirror before leaving for school this morning? If you did, you probably looked at your reflection in a plane mirror. A plane mirror is a mirror with a flat reflecting surface. In a plane mirror, your image looks much the same as it would in a photograph except that your left side is on the right side of your image and your right side is on the left side of your image. Also, your image seems to be coming from behind the mirror. How does a plane mirror form an image?

✔ **Reading Check** *What is a plane mirror?*

Figure 11 shows a person looking into a plane mirror. Light waves from the Sun or another source of light strike each part of the person. These light rays bounce off of the person according to the law of reflection, and some of them strike the mirror. The rays that strike the mirror also are reflected according to the law of reflection. **Figure 11A** shows the path traveled by a few of the rays that have been reflected off the person and reflected back to the person's eye by the mirror.

The Image in a Plane Mirror Why does the image you see in a plane mirror seem to be behind the mirror? This is a result of how your brain processes the light rays that enter your eyes. Although the light rays bounced off the mirror's surface, your brain interprets them as having followed the path shown by the dashed lines in **Figure 11B.** In other words, your brain always assumes that light rays travel in straight lines without changing direction. This makes the reflected light rays look as if they are coming from behind the mirror, even though no source of light is there. The image also seems to be the same distance behind the mirror as the person is in front of the mirror.

Figure 11
A plane mirror forms an image by changing the direction of light rays. **A** Light rays that bounce off of a person strike the mirror. Some of these light rays are reflected into the person's eye. **B** The light rays that are shown entering the person's eye seem to be coming from a person behind the mirror.

Physics
INTEGRATION

When a particle like a marble or a basketball bounces off a surface, it obeys the law of reflection. Because light also obeys the law of reflection, people once thought that light must be a stream of particles. Today, experiments have shown that light can behave as though it were both a wave and a stream of energy bundles called photons. Read an article about photons and write a description in your Science Journal.

Concave and Convex Mirrors

Some mirrors are not flat. A concave mirror has a surface that is curved inward, like the inside of a spoon. Unlike plane mirrors, concave mirrors cause light rays to come together, or converge. A convex mirror, on the other hand, has a surface that curves outward, like the outside of a spoon. Convex mirrors cause light waves to spread out, or diverge. These two types of mirrors form images that are different from the images that are formed by plane mirrors. Examples of a concave and a convex mirror are shown in **Figure 12.**

✔ **Reading Check** *What's the difference between a concave and convex mirror?*

Concave Mirrors The way in which a concave mirror forms an image is shown in **Figure 13.** A straight line drawn perpendicular to the center of a concave or convex mirror is called the optical axis. Light rays that travel parallel to the optical axis and strike the mirror are reflected so that they pass through a single point on the optical axis called the **focal point.** The distance along the optical axis from the center of the mirror to the focal point is called the **focal length.**

The image formed by a concave mirror depends on the position of the object relative to its focal point. If the object is farther from the mirror than the focal point, the image appears to be upside down, or inverted. The size of the image decreases as the object is moved farther away from the mirror. If the object is closer to the mirror than one focal length, the image is upright and gets larger as the object moves closer to the mirror.

A concave mirror can produce a focused beam of light if a source of light is placed at the mirror's focal point, as shown in **Figure 13.** Flashlights and automobile headlights use concave mirrors to produce directed beams of light.

Figure 12
Not all mirrors are flat.
A A concave mirror has a surface that's curved inward. **B** A convex mirror has a surface that's curved outward.

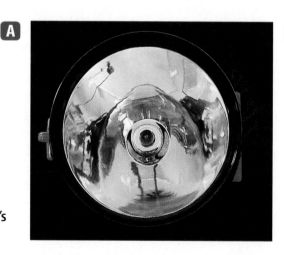
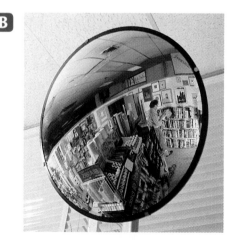

Figure 13

Glance into a flat plane mirror and you'll see an upright, same-size image of yourself. But look into a concave mirror, and you'll see yourself larger than life, right side up, or upside down—or not at all! This is because the way a concave mirror forms an image depends on the position of an object in front of the mirror, as shown here.

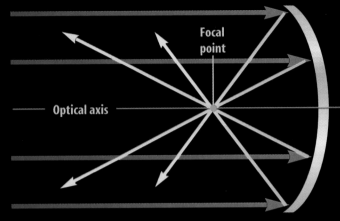

A concave mirror reflects all light rays traveling parallel to the optical axis so that they pass through the focal point.

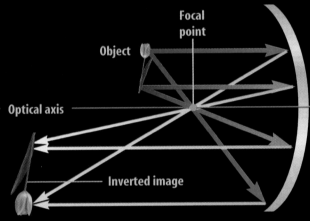

When an object, such as this flower, is placed beyond the focal point, the mirror forms an image that is inverted.

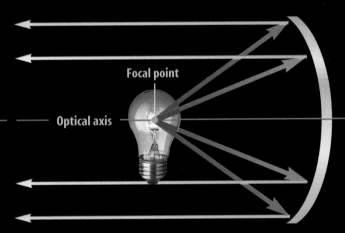

When a source of light is placed at the focal point, a beam of parallel light rays is formed. The concave mirror in a flashlight, for example, creates a concentrated beam of parallel light rays.

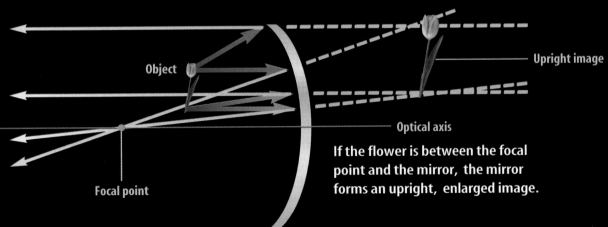

If the flower is between the focal point and the mirror, the mirror forms an upright, enlarged image.

Forming an Image with a Lens

Procedure

1. Fill a **glass test tube** with **water** and seal it with a **stopper.**
2. Write your name on a **10-cm × 10-cm card.** Lay the test tube on the card and observe the appearance of your name.
3. Hold the test tube about 1 cm above the card and observe the appearance of your name. Record your observations.
4. Observe what happens to your name as you slowly move the test tube away from the card. Record your observations.

Analysis

1. Is the water-filled test tube a concave lens or a convex lens?
2. Compare the image that formed when the test tube was close to the card with the image that formed when the test tube was far from the card.

Telescopes

Just as microscopes are used to magnify very small objects, telescopes are used to examine objects that are very far away. The first telescopes were made at about the same time as the first microscopes. Much of what we know about the Moon, the solar system, and the distant universe has come from images and other information gathered by telescopes.

Refracting Telescopes The simplest **refracting telescopes** use two convex lenses to form an image of a distant object. Just as in a compound microscope, light passes through an objective lens that forms an image. That image is then magnified by an eyepiece, as shown in **Figure 23A.**

An important difference between a telescope and a microscope is the size of the objective lens. The main purpose of a telescope is not to magnify an image. A telescope's main purpose is to gather as much light as possible from distant objects. The larger an objective lens is, the more light that can enter it. This makes images of faraway objects look brighter and more detailed when they are magnified by the eyepiece. With a large enough objective lens, it's possible to see stars and galaxies that are many trillions of kilometers away. **Figure 23B** shows the largest refracting telescope ever made.

✔ **Reading Check** *How does a telescope's objective lens enable distant objects to be seen?*

B The refracting telescope at the Yerkes Observatory in Wisconsin has the largest objective lens in the world. It has a diameter of 1 m.

Figure 23
Refracting telescopes use two convex lenses.

Objective lens

A A refracting telescope is made from an objective lens and an eyepiece. The objective lens forms an image that is magnified by the eyepiece.

Eyepiece lens

Figure 24
Reflecting telescopes gather light by using a concave mirror.

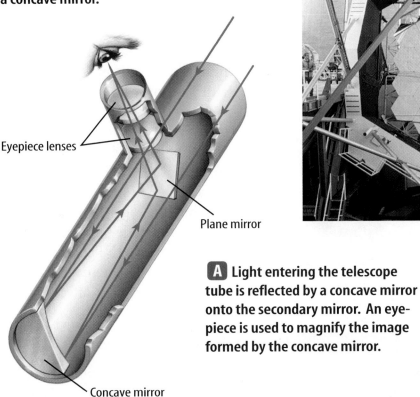

Eyepiece lenses

Plane mirror

Concave mirror

A Light entering the telescope tube is reflected by a concave mirror onto the secondary mirror. An eyepiece is used to magnify the image formed by the concave mirror.

B The concave mirror of the Keck telescope in Mauna Kea, Hawaii, is made of six-sided segments that are 1.8 m across.

Reflecting Telescopes Refracting telescopes have size limitations. One problem is that the objective lens can be supported only around its edges. If the lens is extremely large, it cannot be supported enough to keep the glass from sagging slightly under its own weight. This causes the image that the lens forms to become distorted.

Reflecting telescopes can be made much larger than refracting telescopes. **Reflecting telescopes** have a concave mirror instead of a concave objective lens to gather the light from distant objects. As shown in **Figure 24A,** the large concave mirror focuses light onto a secondary mirror that directs it to the eyepiece, which magnifies the image.

Because only the one reflecting surface on the mirror needs to be made carefully and kept clean, telescope mirrors are less expensive to make and maintain than lenses of a similar size. Also, mirrors can be supported not only at their edges but also on their back sides. They can be made much larger without sagging under their own weight. The Keck telescope in Hawaii, shown in **Figure 24B,** is the largest reflecting telescope in the world. Its large concave mirror is 10 m in diameter, and is made of 36 six-sided segments. Each segment is 1.8 m in size and the segments are pieced together to form the mirror.

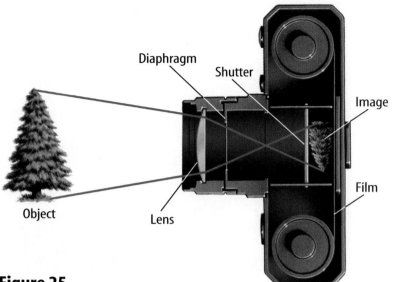

Diaphragm

Shutter

Image

Film

Object

Lens

Figure 25
A camera uses a convex lens to form an image on a piece of light-sensitive film. The image formed by a camera lens is smaller than the object.

Cameras

You probably see photographs taken by cameras almost every day. A typical camera uses a convex lens to form an image on a section of film, just as your eye's lens focuses an image on your retina. The convex lens has a short focal length so that it forms an image that is smaller than the object and inverted on the film. Look at the camera shown in **Figure 25.** When the shutter is open, the convex lens focuses an image on a piece of film that is sensitive to light. Light-sensitive film contains chemicals that undergo chemical reactions when light hits it. The brighter parts of the image affect the film more than the darker parts do.

☑ **Reading Check** *What type of lens does a camera use?*

An image that is too bright can overexpose the film by causing too much of the light-sensitive material to react. On the other hand, if too little light reaches the film, the image might be too dark. To control how much light reaches the film, many cameras have a device called a diaphragm. The diaphragm is opened to let more light onto the film and closed to reduce the amount of light that strikes the film.

Lasers

Perhaps you've seen the narrow, intense beams of laser light used in a laser light show. Intense laser beams are also used for different kinds of surgery. Why can laser beams be so intense? One reason is that, unlike ordinary light, a laser beam doesn't spread out as it travels.

Spreading Light Beams Suppose you shine a flashlight on a wall in a darkened room. The size of the spot of light on the wall depends on the distance between the flashlight and the wall. As the flashlight moves farther from the wall, the spot of light gets larger. This is because the beam of light produced by the flashlight spreads out as it travels. As a result, the energy carried by the light waves in the beam is spread over an increasingly larger area as the light beam travels. As the energy is spread over a larger area, the energy becomes less concentrated and the intensity of the beam decreases.

Figure 26
Laser light is different from the light produced by a lightbulb.

A The light from a bulb contains waves with many different wavelengths that are out of phase and traveling in different directions.

B The light from a laser contains waves with only one wavelength that are in phase and traveling in the same direction.

Using Laser Light Laser light is different from the light produced by the flashlight in several ways, as shown in **Figure 26.** One difference is that in a beam of laser light, the crests and troughs of the light waves overlap, so the waves are in phase.

Because a laser beam doesn't spread out, a large amount of energy can be applied to a very small area. This property enables lasers to be used for cutting and welding materials and as a replacement for scalpels in surgery. Less intense laser light is used for such applications as reading and writing to CDs or in grocery store bar-code readers. Surveyors and builders use lasers to measure distances, angles, and heights. Laser beams also are used to transmit information through space or through optical fibers.

Section Assessment

1. How is a compound microscope different from a magnifying lens?

2. Compare and contrast reflecting and refracting telescope. Why aren't refracting telescopes bigger than reflecting telescopes?

3. Why is the objective lens of a refracting telescope bigger than the objective lens of a microscope?

4. Describe how laser light is different from the light produced by a light bulb.

5. **Think Critically** Could a camera with a concave lens instead of a convex lens still take pictures? Explain.

Skill Builder Activities

6. **Communicating** Using words, pictures, or other media, think of a way to explain to a friend how convex and concave lenses work. **For more help,** refer to the Science Skill Handbook.

7. **Solving One-Step Equations** The size of an image is related to the magnification of an optical instrument by the following formula:

    Image size = magnification × object size

    A blood cell has a diameter of about 0.001 cm. How large is the image formed by a microscope with a magnification of 1,000? **For more help,** refer to the Math Skill Handbook.

Activity

Image Formation by a Convex Lens

The type of image formed by a convex lens, also called a converging lens, is related to the distance of the object from the lens. This distance is called the object distance. The location of the image also is related to the distance of the object from the lens. The distance from the lens to the image is called the image distance. What happens to the position of the image as the object gets nearer or farther from the lens?

What You'll Investigate

How are the image distance and object distance related for a convex lens?

Materials

convex lens
modeling clay
meterstick
flashlight
masking tape
20-cm square piece of cardboard
with a white surface

Goals

- **Measure** the image distance as the object distance changes.
- **Observe** the type of image formed as the object distance changes.

Safety Precautions

Procedure

1. **Design** a data table to record your data in. Make three columns in your table— one column for the object distance, another for the image distance, and the third for the type of image.

2. Use the modeling clay to make the lens stand upright on the lab table.

3. Form the letter *F* on the glass surface of the flashlight with masking tape.

4. Turn on the flashlight and place it 1 m from the lens. Position the flashlight so the flashlight beam is shining through the lens.

5. **Record** the distance from the flashlight to the lens in the object distance column in your data table.

6. Hold the cardboard vertically upright on the other side of the lens, and move it back and forth until a sharp image of the letter *F* is obtained.

Convex Lens Data		
Object Distance (m)	Image Distance (m)	Image Type
1.00	0.43	inverted, smaller
0.50	0.75	inverted, larger
0.25	1.50	upright, larger

7. **Measure** the distance of the card from the lens using the meterstick, and record this distance in the image distance column in your data table.

8. **Record** in the third column of your data table whether the image is upright or inverted, and smaller or larger.

9. Repeat steps 4 through 8 for object distances of 0.50 m and 0.25 m and record your data in your data table.

Conclude and Apply

1. How did the image distance change as the object distance decreased?

2. How did the image change as the object distance decreased?

3. What would happen to the size of the image if the flashlight were much farther away than 1 m?

*C*ommunicating
Your Data

Demonstrate this activity to a third-grade class and explain how it works. **For more help, refer to the** Science Skill Handbook.

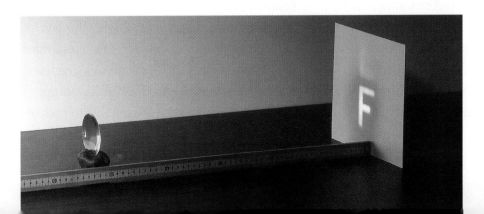

Eyeglasses

"**I**t is not yet twenty years since the art of making spectacles, one of the most useful arts on Earth, was discovered. I myself have seen and conversed with the man who made them first."

This quote from an Italian monk dates back to 1306 and is one of the first historical records to refer to eyeglasses. Unfortunately, the monk, Giordano, never actually named the man he met. Thus, the inventor of eyeglasses—one of the most widely used forms of technology today—remains unknown.

The mystery exists, in part, because different cultures in different places used some type of magnifying tool to improve their vision. These tools eventually merged into what today is recognized as a pair of glasses. For example, a rock-crystal lens made by early Assyrians who lived 3,500 years ago in what is now Iraq, may have been used to improve vision. About 2,000 years ago, the Roman writer Seneca looked through a glass globe of water to make the letters appear bigger in the books he read. By the 10th century, glasses were invented in China, but they were used to keep away bad luck, not to improve vision. Trade between China and Europe, however, likely led some unknown inventor to come up with an idea.

The inventor fused two metal-ringed magnifying lenses together so they could perch on the nose.

In 1456, the printing press was invented. Suddenly, there was more to read, which, in turn, made the ability to see clearly more important. In Europe, eyeglasses began to appear in paintings of scholars, clergy, and the upper classes—the only people who knew how to read at the time. Although the ability to read spread fairly quickly, eyeglasses were so expensive only the rich could afford them. In the early 1700s, for example, glasses cost roughly $200, which is comparable to thousands of dollars today. By the mid-1800s, improvements in manufacturing techniques made eyeglasses much less expensive to make, and thus this important invention became widely available to people of all walks of life.

Cheryl Landry at work with Bosnian teenage soldier

Inventor Unknown

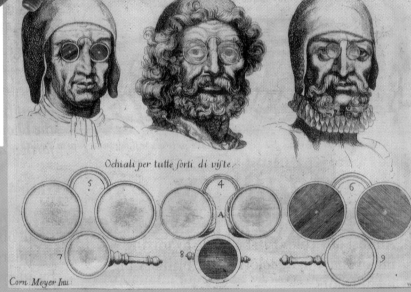

Ochiali per tutte forti di vifte

Corn. Meyer Inu.

This Italian engraving from the 1600s shows glasses of all strengths.

How Eyeglasses Work

Eyeglasses are used to correct farsightedness and nearsightedness, as well as other vision problems. Farsighted people have difficulty seeing things close up because light rays from nearby objects do not converge enough to form an image on the retina. This problem can be corrected by using convex lenses that cause light rays to converge before they enter the eye. Nearsighted people have problems seeing distant objects because light rays from far-away objects are focused in front of the retina. Concave lenses that cause light rays to diverge are used to correct this vision problem.

CONNECTIONS **Research** In many parts of the world, people have no vision care, and eye diseases and poor vision go untreated. Research the work of groups that bring eye care to people. Start with eye doctor Cheryl Landry, who works with the Bosnian Children's Fund.

SCIENCE *Online*

For more information, visit
science.glencoe.com

Reading Comprehension

Read the passage. Then read each question that follows the passage. Decide which is the best answer to each question.

The History of the Telescope: An International Story

Roger Bacon, an English scientist, first wrote about the basic ideas behind the operation of a telescope in the 1200s. It was not until the early 1600s, however that Han Lippershey, a Dutchman who made spectacles for people with poor vision, made the first telescope. Lippershey noticed that objects appeared closer if he viewed them through a combination of a concave and a convex lens. He placed the lenses in a tube to hold them more easily. This was the world's first refracting telescope.

A few years later, an Italian scientist, Galileo, was the first to point a telescope toward the stars. Galileo learned of the Dutch invention in 1609. At the time, it was mainly used to see objects on Earth, such as distant ships and enemy armies. This is why the telescope was first called a "spyglass." Galileo made his own telescope and began using it to view the sky. Before this, Galileo had not been particularly interested in astronomy. That quickly changed as he recorded observations of the Moon's surface, spots on the Sun, and four moons circling Jupiter.

Another advance in telescope technology occurred in 1663 when James Gregory, a Scottish scientist, designed the first reflecting telescope. Isaac Newton built the first reflecting telescope 25 years later. The earliest, most valuable contribution to astronomy made by an American was the construction of the Hooker telescope, a reflecting telescope on Mount Wilson. Completed in 1917, its 254 cm reflecting concave mirror allowed astronomers to see other galaxies clearly for the first time.

Since then, scientists have continued to design and build larger and more powerful telescopes. The development of the modern telescope is the result of many years of work by many scientists across the world.

Test-Taking Tip As you read the passage, make a time line of the history of the telescope.

1. The telescope was first called a "spyglass" because it _____.
   A) was helpful in observing the Moon and stars
   B) was designed by Roger Bacon
   C) could be used to watch other people
   D) was first made by a Dutchman

2. According to the passage, scientists often _____.
   F) build upon one other's work
   G) are slow workers
   H) aren't interested in many things
   J) never read the work of other scientists

3. The earliest, most valuable contribution to astronomy made by an American was _____.
   A) the first refracting telescope built in the 1600s
   B) Roger Bacon's basic ideas about the operation of a telescope in the 1200s
   C) the construction of the Hooker telescope on Mount Wilson which allowed astronomers to see other galaxies clearly for the first time
   D) using a telescope to view the Moon's surface, spots on the Sun, and four moons circling Jupiter

Reasoning and Skills

Power of a Lens		
Lens	Diopter	Focal length (m)
1	1/4	4
2	1/5	5
3	1/6	6
4	1/7	7
5	1/9	?

1. Diopters are one way to measure the strength of a lens. What is the focal length of lens 5?
- **A)** 5
- **B)** 8
- **C)** 9
- **D)** 10

Test-Taking Tip Study the values for the first three lenses and consider how the diopter value is related to the focal length.

Wavelengths of Electromagnetic Waves	
Type of Wave	Wavelength
radio wave	>0.1 m
microwave	0.001m – 0.1 m
infrared wave	0.000 000 7 – 0.001 m
visible light wave	0.000 000 4 – 0.000 000 7 m

2. A wave with a wavelength of 0.03 m would be what type of wave?
- **F)** radio wave
- **G)** microwave
- **H)** infrared wave
- **J)** visible light wave

Test-Taking Tip Read the table's column headings carefully and then reread the question.

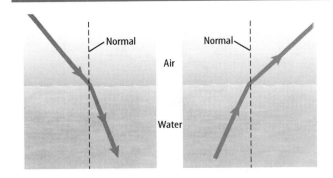

3. What is happening to the wave in this figure?
- **A)** it is being diffracted
- **B)** it is being refracted
- **C)** it is experiencing constructive interference
- **D)** it is experiencing destructive interference

Test-Taking Tip Review the difference between diffraction, refraction, constructive interference, and destructive interference.

4. Explain the relationship between the amplitude of a sound wave and the loudness of a sound. How is the amplitude of the sound wave related to the amount of energy it carries?

Test-Taking Tip Recall the definition of the terms *amplitude* and *loudness*.

Student Resources

Student Resources

CONTENTS

Field GUIDE

Musical Instruments

Some people have defined music as "patterns of tones." A tone is a sound with a specific pitch. In music, a tone might also be called a note. Pitch describes how high or low the tone is. Like all sounds, musical tones are produced when an object vibrates. Higher pitches are produced by more vibrations per second, and therefore have a higher frequency.

Most musical instruments use resonance to amplify sounds. To amplify a sound means to increase its volume. Resonance occurs when one object causes another object to vibrate at the same frequency—or pitch.

How Resonance Works

A vibrating object produces sound waves. These waves can affect other objects and cause them to vibrate. As more matter vibrates, a louder sound is produced. For example, resonance is at work in a guitar. When a guitar's strings are plucked or strummed, the strings vibrate. The strings' vibrations make the thin soundboard—in this case, the front of the guitar—vibrate. The soundboard's vibrations make the air inside the guitar's hollow body vibrate. The vibrating air amplifies the sounds that were first produced by the strings.

Stringed Instruments

Tones are produced in stringed instruments by making stretched strings vibrate. Each string is tuned to a different pitch. When playing stringed instruments such as the harp, each string produces only one pitch. The player creates different pitches by plucking different strings.

When playing stringed instruments such as the guitar and violin, the player can change the pitch of each string by pressing down on one end and making it shorter. Stringed instruments may be strummed, plucked, or played with a bow.

Mandolin

Harp

Field Activity

Watch an orchestra or band perform in a live concert or on television. In your Science Journal, name all of the different instruments you recognize. Then use this field guide to identify the category in which each instrument is classified.

Wind Instruments—Woodwinds

Woodwind instruments include the clarinet, the saxophone, and the recorder. These instruments are played by blowing into a mouthpiece or across a hole. Some woodwinds, such as clarinets, have a thin flexible reed in the mouthpiece that vibrates. The reed causes the air in the tube to vibrate. As the air vibrates inside the woodwind's hollow tube, tones are produced. Musicians change this instrument's pitch by covering holes with their fingers or by pressing keys that cover holes. Covering a hole lengthens or shortens the column of air inside the tube.

Saxophone

Clarinet

Trombone

Tuba

Wind Instruments—Brass

Brass instruments include the trumpet, the trombone, and the tuba. Their mouthpieces are larger than woodwinds' mouthpieces. Brass instruments are played by pressing the lips against a mouthpiece and blowing so the lips vibrate. Musicians change brass instruments' pitches by tensing or relaxing their lips. With most brass instruments, the pitch also can be changed by pressing valves, which changes the length of the vibrating column of air inside the instrument.

Percussion Instruments—Idiophones

Idiophones vibrate to produce tones. Musicians play them by hitting, shaking, scraping, or plucking them. Idiophones such as cymbals, bells, gongs, music boxes, and xylophone keys play only one pitch. Triangles, clappers, rattles, and cymbals have indefinite pitches—their pitches depend on how they are played and how they are constructed.

Xylophone

Percussion Instruments—Membranophones

Membranophones produce sound when their membranes—the stretched tops of drums or the tiny membranes within kazoos—vibrate. Drums are usually struck with hands, with beaters such as drumsticks, or with knotted cords to produce tones.

Bongo

Electric Instruments

Electric instruments such as the electric guitar and electric violin are played like regular instruments. However, rather than using resonance to amplify their sound, their vibrations are converted to electrical signals that are amplified electronically. The amplified electric signal is then converted to sound by a loudspeaker.

Guitar and Amplifier

Keyboard Instruments—Piano

Each piano key is attached to a small hammer. When the player presses a key, the hammer hits a string and makes it vibrate. The strings are different lengths and each string produces a different pitch. The piano's body amplifies the tones.

Piano

Keyboards—Pipe Organ

Pressing a pipe organ's key opens a pipe to let air vibrate inside it. The pipes are different lengths, and each produces a different pitch.

Pipe Organ

Electronic Instruments

Unlike all other types of musical instruments, electronic instruments do not rely on vibrations to produce sounds. Instead, these instruments produce electrical signals that a computer then converts to sounds. Even though a synthesizer has a keyboard, it is classified as an electronic instrument because it produces sounds electronically. Today, it is the most widely used electronic instrument.

As you study science, you will make many observations and conduct investigations and experiments. You will also research information that is available from many sources. These activities will involve organizing and recording data. The quality of the data you collect and the way you organize it will determine how well others can understand and use it. In **Figure 1,** the student is obtaining and recording information using a thermometer.

Putting your observations in writing is an important way of communicating to others the information you have found and the results of your investigations and experiments.

Researching Information

Scientists work to build on and add to human knowledge of the world. Before moving in a new direction, it is important to gather the information that already is known about a subject. You will look for such information in various reference sources. Follow these steps to research information on a scientific subject:

Step 1 Determine exactly what you need to know about the subject. For instance, you might want to find out about one of the elements in the periodic table.

Step 2 Make a list of questions, such as: Who discovered the element? When was it discovered? What makes the element useful or interesting?

Step 3 Use multiple sources such as textbooks, encyclopedias, government documents, professional journals, science magazines, and the Internet.

Step 4 List where you found the sources. Make sure the sources you use are reliable and the most current available.

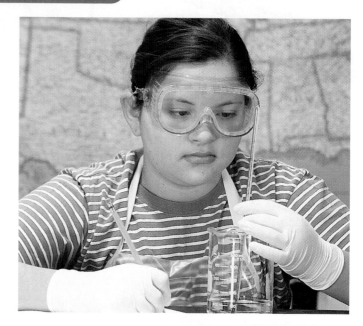

Figure 1
Making an observation is one way to gather information directly.

Evaluating Print and Nonprint Sources

Not all sources of information are reliable. Evaluate the sources you use for information, and use only those you know to be dependable. For example, suppose you want to find ways to make your home more energy efficient. You might find two Web sites on how to save energy in your home. One Web site contains "Energy-Saving Tips" written by a company that sells a new type of weatherproofing material you put around your door frames. The other is a Web page on "Conserving Energy in Your Home" written by the U.S. Department of Energy. You would choose the second Web site as the more reliable source of information.

In science, information can change rapidly. Always consult the most current sources. A 1985 source about saving energy would not reflect the most recent research and findings.

Interpreting Scientific Illustrations

As you research a science topic, you will see drawings, diagrams, and photographs. Illustrations help you understand what you read. Some illustrations are included to help you understand an idea that you can't see easily by yourself. For instance, you can't see the tiny particles in an atom, but you can look at a diagram of an atom as labeled in **Figure 2** that helps you understand something about it. Visualizing a drawing helps many people remember details more easily. Illustrations also provide examples that clarify difficult concepts or give additional information about the topic you are studying.

Most illustrations have a label or caption. A label or caption identifies the illustration or provides additional information to better explain it. Can you find the caption or labels in **Figure 2?**

Figure 2
This drawing shows an atom of carbon with its six protons, six neutrons, and six electrons.

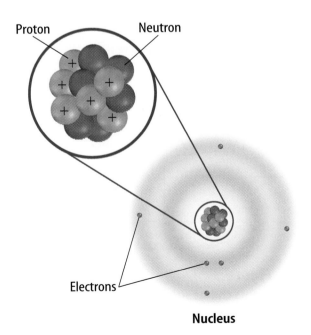

Concept Mapping

If you were taking a car trip, you might take some sort of road map. By using a map, you begin to learn where you are in relation to other places on the map.

A concept map is similar to a road map, but a concept map shows relationships among ideas (or concepts) rather than places. It is a diagram that visually shows how concepts are related. Because a concept map shows relationships among ideas, it can make the meanings of ideas and terms clear and help you understand what you are studying.

Overall, concept maps are useful for breaking large concepts down into smaller parts, making learning easier.

Venn Diagram

Although it is not a concept map, a Venn diagram illustrates how two subjects compare and contrast. In other words, you can see the characteristics that the subjects have in common and those that they do not.

The Venn diagram in **Figure 3** shows the relationship between two different substances made from the element carbon. However, due to the way their atoms are arranged, one substance is the gemstone diamond, and the other is the graphite found in pencils.

Figure 3
A Venn diagram shows how objects or concepts are alike and how they are different.

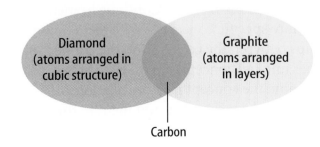

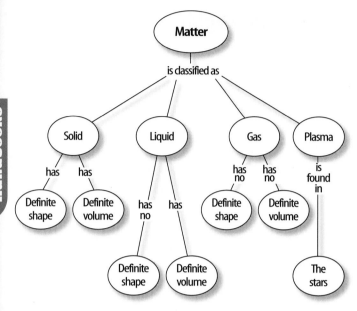

Figure 4
A network tree shows how concepts or objects are related.

Network Tree Look at the concept map in **Figure 4,** that describes the different types of matter. This is called a network tree concept map. Notice how some words are in ovals while others are written across connecting lines. The words inside the ovals are science terms or concepts. The words written on the connecting lines describe the relationships between the concepts.

When constructing a network tree, write the topic on a note card or piece of paper. Write the major concepts related to that topic on separate note cards or pieces of paper. Then arrange them in order from general to specific. Branch the related concepts from the major concept and describe the relationships on the connecting lines. Continue branching to more specific concepts. Write the relationships between the concepts on the connecting lines until all concepts are mapped. Then examine the concept map for relationships that cross branches, and add them to the concept map.

Events Chain An events chain is another type of concept map. It models the order of items or their sequence. In science, an events chain can be used to describe a sequence of events, the steps in a procedure, or the stages of a process.

When making an events chain, first find the one event that starts the chain. This event is called the *initiating event.* Then, find the next event in the chain and continue until you reach an outcome. Suppose you are asked to describe why and how a sound might make an echo. You might draw an events chain such as the one in **Figure 5.** Notice that connecting words are not necessary in an events chain.

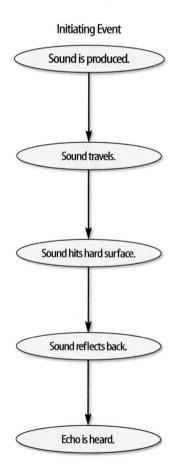

Figure 5
Events chains show the order of steps in a process or event.

Cycle Map A cycle concept map is a specific type of events chain map. In a cycle concept map, the series of events does not produce a final outcome. Instead, the last event in the chain relates back to the beginning event.

You first decide what event will be used as the beginning event. Once that is decided, you list events in order that occur after it. Words are written between events that describe what happens from one event to the next. The last event in a cycle concept map relates back to the beginning event. The number of events in a cycle concept varies, but is usually three or more. Look at the cycle map, as shown in **Figure 6.**

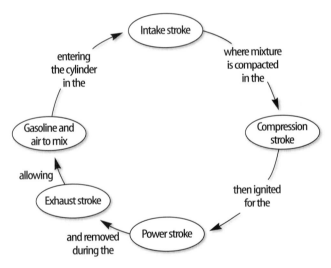

Figure 6
A cycle map shows events that occur in a cycle.

Spider Map A type of concept map that you can use for brainstorming is the spider map. When you have a central idea, you might find you have a jumble of ideas that relate to it but are not necessarily clearly related to each other. The spider map on sound in **Figure 7** shows that if you write these ideas outside the main concept, then you can begin to separate and group un-related terms so they become more useful.

Figure 7
A spider map allows you to list ideas that relate to a central topic but not necessarily to one another.

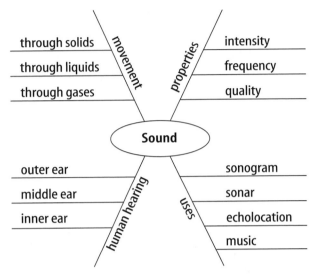

Writing a Paper

You will write papers often when researching science topics or reporting the results of investigations or experiments. Scientists frequently write papers to share their data and conclusions with other scientists and the public. When writing a paper, use these steps.

Step 1 Assemble your data by using graphs, tables, or a concept map. Create an outline.

Step 2 Start with an introduction that contains a clear statement of purpose and what you intend to discuss or prove.

Step 3 Organize the body into paragraphs. Each paragraph should start with a topic sentence, and the remaining sentences in that paragraph should support your point.

Step 4 Position data to help support your points.

Step 5 Summarize the main points and finish with a conclusion statement.

Step 6 Use tables, graphs, charts, and illustrations whenever possible.

Investigating and Experimenting

You might say the work of a scientist is to solve problems. When you decide to find out why your neighbor's hydrangeas produce blue flowers while yours are pink, you are problem solving, too. You might also observe that your neighbor's azaleas are healthier than yours are and decide to see whether differences in the soil explain the differences in these plants.

Scientists use orderly approaches to solve problems. The methods scientists use include identifying a question, making observations, forming a hypothesis, testing a hypothesis, analyzing results, and drawing conclusions.

Scientific investigations involve careful observation under controlled conditions. Such observation of an object or a process can suggest new and interesting questions about it. These questions sometimes lead to the formation of a hypothesis. Scientific investigations are designed to test a hypothesis.

Identifying a Question

The first step in a scientific investigation or experiment is to identify a question to be answered or a problem to be solved. You might be interested in knowing how beams of laser light like the ones in **Figure 8** look the way they do.

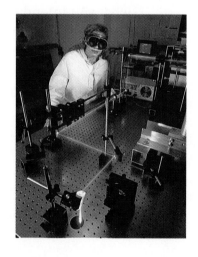

Figure 8
When you see lasers being used for scientific research, you might ask yourself, "Are these lasers different from those that are used for surgery?"

Forming Hypotheses

Hypotheses are based on observations that have been made. A hypothesis is a possible explanation based on previous knowledge and observations.

Perhaps a scientist has observed that certain substances dissolve faster in warm water than in cold. Based on these observations, the scientist can make a statement that he or she can test. The statement is a hypothesis. The hypothesis could be: *A substance dissolves in warm water faster.* A hypothesis has to be something you can test by using an investigation. A testable hypothesis is a valid hypothesis.

Predicting

When you apply a hypothesis, or general explanation, to a specific situation, you predict something about that situation. First, you must identify which hypothesis fits the situation you are considering. People use predictions to make everyday decisions. Based on previous observations and experiences, you might form a prediction that if substances dissolve in warm water faster, then heating the water will shorten mixing time for powdered fruit drinks. Someone could use this prediction to save time in preparing a fruit punch for a party.

Testing a Hypothesis

To test a hypothesis, you need a procedure. A procedure is the plan you follow in your experiment. A procedure tells you what materials to use, as well as how and in what order to use them. When you follow a procedure, data are generated that support or do not support the original hypothesis statement.

For example, premium gasoline costs more than regular gasoline. Does premium gasoline increase the efficiency or fuel mileage of your family car? You decide to test the hypothesis: "If premium gasoline is more efficient, then it should increase the fuel mileage of my family's car." Then you write the procedure shown in **Figure 9** for your experiment and generate the data presented in the table below.

Figure 9
A procedure tells you what to do step by step.

> ## Procedure
> 1. Use regular gasoline for two weeks.
> 2. Record the number of kilometers between fill-ups and the amount of gasoline used.
> 3. Switch to premium gasoline for two weeks.
> 4. Record the number of kilometers between fill-ups and the amount of gasoline used.

Gasoline Data			
Type of Gasoline	Kilometers Traveled	Liters Used	Liters per Kilometer
Regular	762	45.34	0.059
Premium	661	42.30	0.064

These data show that premium gasoline is less efficient than regular gasoline in one particular car. It took more gasoline to travel 1 km (0.064) using premium gasoline than it did to travel 1 km using regular gasoline (0.059). This conclusion does not support the hypothesis.

Are all investigations alike? Keep in mind as you perform investigations in science that a hypothesis can be tested in many ways. Not every investigation makes use of all the ways that are described on these pages, and not all hypotheses are tested by investigations. Scientists encounter many variations in the methods that are used when they perform experiments. The skills in this handbook are here for you to use and practice.

Identifying and Manipulating Variables and Controls

In any experiment, it is important to keep everything the same except for the item you are testing. The one factor you change is called the independent variable. The factor that changes as a result of the independent variable is called the dependent variable. Always make sure you have only one independent variable. If you allow more than one, you will not know what causes the changes you observe in the dependent variable. Many experiments also have controls—individual instances or experimental subjects for which the independent variable is not changed. You can then compare the test results to the control results.

For example, in the fuel-mileage experiment, you made everything the same except the type of gasoline that was used. The driver, the type of automobile, and the type of driving were the same throughout. In this way, you could be sure that any mileage differences were caused by the type of fuel—the independent variable. The fuel mileage was the dependent variable.

If you could repeat the experiment using several automobiles of the same type on a standard driving track with the same driver, you could make one automobile a control by using regular gasoline over the four-week period.

Collecting Data

Whether you are carrying out an investigation or a short observational experiment, you will collect data, or information. Scientists collect data accurately as numbers and descriptions and organize it in specific ways.

Observing Scientists observe items and events, then record what they see. When they use only words to describe an observation, it is called qualitative data. For example, a scientist might describe the color, texture, or odor of a substance produced in a chemical reaction. Scientists' observations also can describe how much there is of something. These observations use numbers, as well as words, in the description and are called quantitative data. For example, if a sample of the element gold is described as being "shiny and very dense," the data are clearly qualitative. Quantitative data on this sample of gold might include "a mass of 30 g and a density of 19.3 g/cm^3." Quantitative data often are organized into tables. Then, from information in the table, a graph can be drawn. Graphs can reveal relationships that exist in experimental data.

When you make observations in science, you should examine the entire object or situation first, then look carefully for details. If you're looking at an element sample, for instance, check the general color and pattern of the sample before using a hand lens to examine its surface for any smaller details or characteristics. Remember to record accurately everything you see.

Scientists try to make careful and accurate observations. When possible, they use instruments such as microscopes, metric rulers, graduated cylinders, thermometers, and balances. Measurements provide numerical data that can be repeated and checked.

Sampling When working with large numbers of objects or a large population, scientists usually cannot observe or study every one of them. Instead, they use a sample or a portion of the total number. To *sample* is to take a small, representative portion of the objects or organisms of a population for research. By making careful observations or manipulating variables within a portion of a group, information is discovered and conclusions are drawn that might apply to the whole population.

Estimating Scientific work also involves estimating. To estimate is to make a judgment about the size or the number of something without measuring or counting every object or member of a population. Scientists first measure or count the amount or number in a small sample. A geologist, for example, might remove a 10-g sample from a large rock that is rich in copper ore, as in **Figure 10.** Then a chemist would determine the percentage of copper by mass and multiply that percentage by the total mass of the rock to estimate the total mass of copper in the large rock.

Figure 10
Determining the percentage of copper by mass that is present in a small piece of a large rock, which is rich in copper ore, can help estimate the total mass of copper ore that is present in the rock.

Measuring in SI

The metric system of measurement was developed in 1795. A modern form of the metric system, called the International System, or SI, was adopted in 1960. SI provides standard measurements that all scientists around the world can understand.

The metric system is convenient because unit sizes vary by multiples of 10. When changing from smaller units to larger units, divide by a multiple of 10. When changing from larger units to smaller, multiply by a multiple of 10. To convert millimeters to centimeters, divide the millimeters by 10. To convert 30 mm to centimeters, divide 30 by 10 (30 mm equal 3 cm).

Prefixes are used to name units. Look at the table below for some common metric prefixes and their meanings. Do you see how the prefix *kilo-* attached to the unit *gram* is *kilogram*, or 1,000 g?

Metric Prefixes			
Prefix	**Symbol**	**Meaning**	
kilo-	k	1,000	thousand
hecto-	h	100	hundred
deka-	da	10	ten
deci-	d	0.1	tenth
centi-	c	0.01	hundredth
milli-	m	0.001	thousandth

Now look at the metric ruler shown in **Figure 11.** The centimeter lines are the long, numbered lines, and the shorter lines are millimeter lines.

When using a metric ruler, line up the 0-cm mark with the end of the object being measured, and read the number of the unit where the object ends, in this instance it would be 4.5 cm.

Figure 11
This metric ruler has centimeter and millimeter divisions.

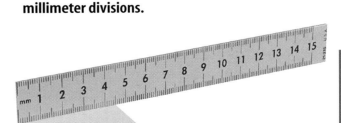

Liquid Volume In some science activities, you will measure liquids. The unit that is used to measure liquids is the liter. A liter has the volume of 1,000 cm³. The prefix *milli-* means "thousandth (0.001)." A milliliter is one thousandth of 1 L, and 1 L has the volume of 1,000 mL. One milliliter of liquid completely fills a cube measuring 1 cm on each side. Therefore, 1 mL equals 1 cm³.

You will use beakers and graduated cylinders to measure liquid volume. A graduated cylinder, as illustrated in **Figure 12,** is marked from bottom to top in milliliters. This one contains 79 mL of a liquid.

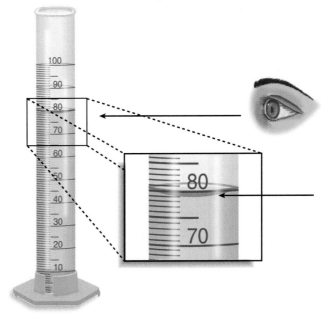

Figure 12
Graduated cylinders measure liquid volume.

Mass Scientists measure mass in grams. You might use a beam balance similar to the one shown in **Figure 13.** The balance has a pan on one side and a set of beams on the other side. Each beam has a rider that slides on the beam.

Before you find the mass of an object, slide all the riders back to the zero point. Check the pointer on the right to make sure it swings an equal distance above and below the zero point. If the swing is unequal, find and turn the adjusting screw until you have an equal swing.

Place an object on the pan. Slide the largest rider along its beam until the pointer drops below zero. Then move it back one notch. Repeat the process on each beam until the pointer swings an equal distance above and below the zero point. Sum the masses on each beam to find the mass of the object. Move all riders back to zero when finished.

Figure 13
A triple beam balance is used to determine the mass of an object.

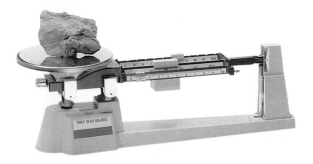

You should never place a hot object on the pan or pour chemicals directly onto the pan. Instead, find the mass of a clean container. Remove the container from the pan, then place the chemicals in the container. Find the mass of the container with the chemicals in it. To find the mass of the chemicals, subtract the mass of the empty container from the mass of the filled container.

Making and Using Tables

Browse through your textbook and you will see tables in the text and in the activities. In a table, data, or information, are arranged so that they are easier to understand. Activity tables help organize the data you collect during an activity so results can be interpreted.

Making Tables To make a table, list the items to be compared in the first column and the characteristics to be compared in the first row. The title should clearly indicate the content of the table, and the column or row heads should tell the reader what information is found in there. The table below lists materials collected for recycling on three weekly pick-up days. The inclusion of kilograms in parentheses also identifies for the reader that the figures are mass units.

Recyclable Materials Collected During Week			
Day of Week	Paper (kg)	Aluminum (kg)	Glass (kg)
Monday	5.0	4.0	12.0
Wednesday	4.0	1.0	10.0
Friday	2.5	2.0	10.0

Using Tables How much paper, in kilograms, is being recycled on Wednesday? Locate the column labeled "Paper (kg)" and the row "Wednesday." The information in the box where the column and row intersect is the answer. Did you answer "4.0"? How much aluminum, in kilograms, is being recycled on Friday? If you answered "2.0," you understand how to read the table. How much glass is collected for recycling each week? Locate the column labeled "Glass (kg)" and add the figures for all three rows. If you answered "32.0," then you know how to locate and use the data provided in the table.

Recording Data

To be useful, the data you collect must be recorded carefully. Accuracy is key. A well-thought-out experiment includes a way to record procedures, observations, and results accurately. Data tables are one way to organize and record results. Set up the tables you will need ahead of time so you can record the data right away.

Record information properly and neatly. Never put unidentified data on scraps of paper. Instead, data should be written in a notebook like the one in **Figure 14.** Write in pencil so information isn't lost if your data gets wet. At each point in the experiment, record your data and label it. That way, your information will be accurate and you will not have to determine what the figures mean when you look at your notes later.

Figure 14
Record data neatly and clearly so it is easy to understand.

Recording Observations

It is important to record observations accurately and completely. That is why you always should record observations in your notes immediately as you make them. It is easy to miss details or make mistakes when recording results from memory. Do not include your personal thoughts when you record your data. Record only what you observe to eliminate bias. For example, when you record the time required for five students to climb the same set of stairs, you would note which student took the longest time. However, you would not refer to that student's time as "the worst time of all the students in the group."

Making Models

You can organize the observations and other data you collect and record in many ways. Making models is one way to help you better understand the parts of a structure you have been observing or the way a process for which you have been taking various measurements works.

Models often show things that are too large or too small for normal viewing. For example, you normally won't see the inside of an atom. However, you can understand the structure of the atom better by making a three-dimensional model of an atom. The relative sizes, the positions, and the movements of protons, neutrons, and electrons can be explained in words. An atomic model made of a plastic-ball nucleus and pipe-cleaner electron shells can help you visualize how the parts of the atom relate to each other.

Other models can be devised on a computer. Some models, such as those that illustrate the chemical combinations of different elements, are mathematical and are represented by equations.

Making and Using Graphs

After scientists organize data in tables, they might display the data in a graph that shows the relationship of one variable to another. A graph makes interpretation and analysis of data easier. Three types of graphs are the line graph, the bar graph, and the circle graph.

Line Graphs A line graph like in **Figure 15** is used to show the relationship between two variables. The variables being compared go on two axes of the graph. For data from an experiment, the independent variable always goes on the horizontal axis, called the *x*-axis. The dependent variable always goes on the vertical axis, called the *y*-axis. After drawing your axes, label each with a scale. Next, plot the data points.

A data point is the intersection of the recorded value of the dependent variable for each tested value of the independent variable. After all the points are plotted, connect them.

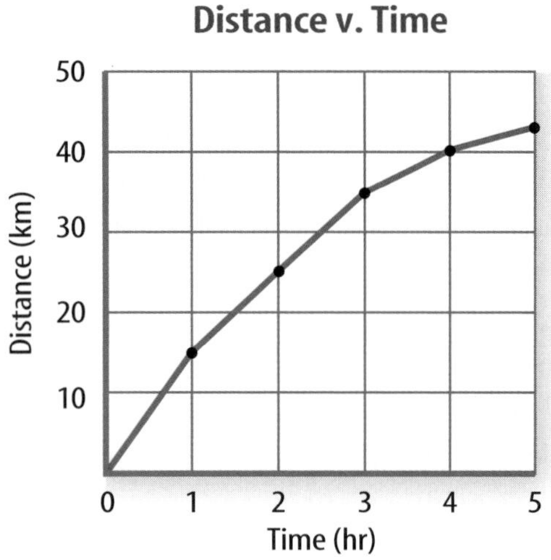

Distance v. Time

(Line graph: Distance (km) on y-axis from 0 to 50, Time (hr) on x-axis from 0 to 5)

Figure 15
This line graph shows the relationship between distance and time during a bicycle ride lasting several hours.

Bar Graphs Bar graphs compare data that do not change continuously. Vertical bars show the relationships among data.

To make a bar graph, set up the *y*-axis as you did for the line graph. Draw vertical bars of equal size from the *x*-axis up to the point on the *y*-axis that represents value of *x*.

Figure 16
The amount of aluminum collected for recycling during one week can be shown as a bar graph or circle graph.

Aluminum Collected During Week

(Bar graph: Mass (kg) on y-axis from 1.0 to 4.0, Day of collection on x-axis showing Monday, Wednesday, Friday)

Circle Graphs A circle graph uses a circle divided into sections to display data as parts (fractions or percentages) of a whole. The size of each section corresponds to the fraction or percentage of the data that the section represents. So, the entire circle represents 100 percent, one-half represents 50 percent, one-fifth represents 20 percent, and so on.

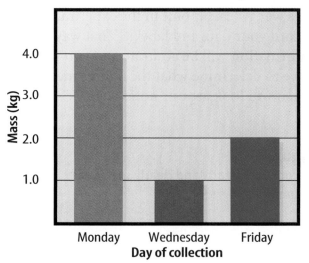

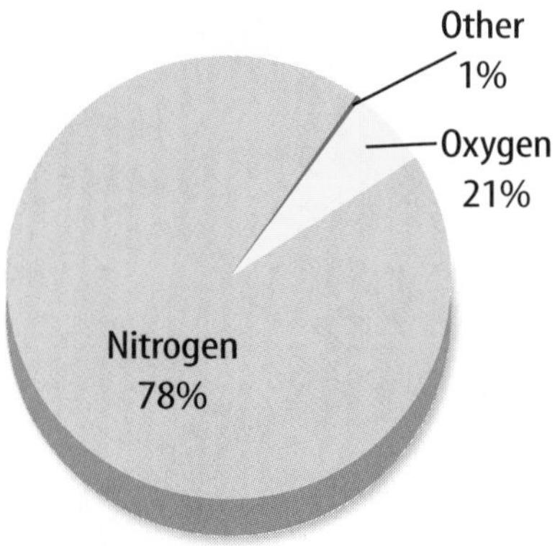

Other 1%

Oxygen 21%

Nitrogen 78%

Analyzing Results

To determine the meaning of your observations and investigation results, you will need to look for patterns in the data. You can organize your information in several of the ways that are discussed in this handbook. Then you must think critically to determine what the data mean. Scientists use several approaches when they analyze the data they have collected and recorded. Each approach is useful for identifying specific patterns in the data.

Forming Operational Definitions

An operational definition defines an object by showing how it functions, works, or behaves. Such definitions are written in terms of how an object works or how it can be used; that is, they describe its job or purpose.

For example, a ruler can be defined as a tool that measures the length of an object (how it can be used). A ruler also can be defined as something that contains a series of marks that can be used as a standard when measuring (how it works).

Classifying

Classifying is the process of sorting objects or events into groups based on common features. When classifying, first observe the objects or events to be classified. Then select one feature that is shared by some members in the group but not by all. Place those members that share that feature into a subgroup. You can classify members into smaller and smaller subgroups based on characteristics.

How might you classify a group of chemicals? You might first classify them by state of matter, putting solids, liquids, and gases into separate groups. Within each group, you could then look for another common feature by which to further classify members of the group, such as color or how reactive they are.

Remember that when you classify, you are grouping objects or events for a purpose. For example, classifying chemicals can be the first step in organizing them for storage. Both at home and at school, poisonous or highly reactive chemicals should all be stored in a safe location where they are not easily accessible to small children or animals. Solids, liquids, and gases each have specific storage requirements that may include waterproof, airtight, or pressurized containers. Are the dangerous chemicals in your home stored in the right place? Keep your purpose in mind as you select the features to form groups and subgroups.

Figure 17
Color is one of many characteristics that are used to classify chemicals.

Science Skill Handbook

Skill Handbooks

Comparing and Contrasting

Observations can be analyzed by noting the similarities and differences between two or more objects or events that you observe. When you look at objects or events to see how they are similar, you are comparing them. Contrasting is looking for differences in objects or events. The table below compares and contrasts the characteristics of two elements.

Elemental Characteristics		
Element	Aluminum	Gold
Color	silver	gold
Classification	metal	metal
Density (g/cm³)	2.7	19.3
Melting Point (°C)	660	1064

Recognizing Cause and Effect

Have you ever heard a loud pop right before the power went out and then suggested that an electric transformer probably blew out? If so, you have observed an effect and inferred a cause. The event is the effect, and the reason for the event is the cause.

When scientists are unsure of the cause of a certain event, they design controlled experiments to determine what caused it.

Interpreting Data

The word *interpret* means "to explain the meaning of something." Look at the problem originally being explored in an experiment and figure out what the data show. Identify the control group and the test group so you can see whether or not changes in the independent variable have had an effect. Look for differences in the dependent variable between the control and test groups.

These differences you observe can be qualitative or quantitative. You would be able to describe a qualitative difference using only words, whereas you would measure a quantitative difference and describe it using numbers. If there are differences, the independent variable that is being tested could have had an effect. If no differences are found between the control and test groups, the variable that is being tested apparently had no effect.

For example, suppose that three beakers each contain 100 mL of water. The beakers are placed on hot plates, and two of the hot plates are turned on, but the third is left off for a period of 5 min. Suppose you are then asked to describe any differences in the water in the three beakers. A qualitative difference might be the appearance of bubbles rising to the top in the water that is being heated but no rising bubbles in the unheated water. A quantitative difference might be a difference in the amount of water that is present in the beakers.

Inferring Scientists often make inferences based on their observations. An inference is an attempt to explain, or interpret, observations or to indicate what caused what you observed. An inference is a type of conclusion.

When making an inference, be certain to use accurate data and accurately described observations. Analyze all of the data that you've collected. Then, based on everything you know, explain or interpret what you've observed.

Drawing Conclusions

When scientists have analyzed the data they collected, they proceed to draw conclusions about what the data mean. These conclusions are sometimes stated using words similar to those found in the hypothesis formed earlier in the process.

Conclusions To analyze your data, you must review all of the observations and measurements that you made and recorded. Recheck all data for accuracy. After your data are rechecked and organized, you are almost ready to draw a conclusion such as "salt water boils at a higher temperature than freshwater."

Before you can draw a conclusion, however, you must determine whether the data allow you to come to a conclusion that supports a hypothesis. Sometimes that will be the case, other times it will not.

If your data do not support a hypothesis, it does not mean that the hypothesis is wrong. It means only that the results of the investigation did not support the hypothesis. Maybe the experiment needs to be redesigned, but very likely, some of the initial observations on which the hypothesis was based were incomplete or biased. Perhaps more observation or research is needed to refine the hypothesis.

Avoiding Bias Sometimes drawing a conclusion involves making judgments. When you make a judgment, you form an opinion about what your data mean. It is important to be honest and to avoid reaching a conclusion if there were no supporting evidence for it or if it were based on a small sample. It also is important not to allow any expectations of results to bias your judgments. If possible, it is a good idea to collect additional data. Scientists do this all the time.

For example, the *Hubble Space Telescope* was sent into space in April, 1990, to provide scientists with clearer views of the universe. The *Hubble* is the size of a school bus and has a 2.4-m-diameter mirror. The *Hubble* helped scientists answer questions about the planet Pluto.

For many years, scientists had only been able to hypothesize about the surface of the planet Pluto. The *Hubble* has now provided pictures of Pluto's surface that show a rough texture with light and dark regions on it. This might be the best information about Pluto scientists will have until they are able to send a space probe to it.

Evaluating Others' Data and Conclusions

Sometimes scientists have to use data that they did not collect themselves, or they have to rely on observations and conclusions drawn by other researchers. In cases such as these, the data must be evaluated carefully.

How were the data obtained? How was the investigation done? Was it carried out properly? Has it been duplicated by other researchers? Were they able to follow the exact procedure? Did they come up with the same results? Look at the conclusion, as well. Would you reach the same conclusion from these results? Only when you have confidence in the data of others can you believe it is true and feel comfortable using it.

Communicating

The communication of ideas is an important part of the work of scientists. A discovery that is not reported will not advance the scientific community's understanding or knowledge. Communication among scientists also is important as a way of improving their investigations.

Scientists communicate in many ways, from writing articles in journals and magazines that explain their investigations and experiments, to announcing important discoveries on television and radio, to sharing ideas with colleagues on the Internet or presenting them as lectures.

Skill Handbooks

Computer Skills

People who study science rely on computers to record and store data and to analyze results from investigations. Whether you work in a laboratory or just need to write a lab report with tables, good computer skills are a necessity.

Using a Word Processor

Suppose your teacher has assigned a written report. After you've completed your research and decided how you want to write the information, you need to put all that information on paper. The easiest way to do this is with a word processing application on a computer.

A computer application that allows you to type your information, change it as many times as you need to, and then print it out so that it looks neat and clean is called a word processing application. You also can use this type of application to create tables and columns, add bullets or cartoon art to your page, include page numbers, and check your spelling.

Helpful Hints

- If you aren't sure how to do something using your word processing program, look in the help menu. You will find a list of topics there to click on for help. After you locate the help topic you need, just follow the step-by-step instructions you see on your screen.
- Just because you've spell checked your report doesn't mean that the spelling is perfect. The spell check feature can't catch misspelled words that look like other words. If you've accidentally typed *cold* instead of *gold*, the spell checker won't know the difference. Always reread your report to make sure you didn't miss any mistakes.

Figure 18
You can use computer programs to make graphs and tables.

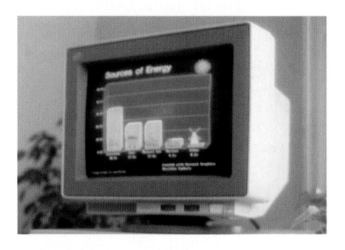

Using a Database

Imagine you're in the middle of a research project, busily gathering facts and information. You soon realize that it's becoming more difficult to organize and keep track of all the information. The tool to use to solve information overload is a database. Just as a file cabinet organizes paper records, a database organizes computer records. However, a database is more powerful than a simple file cabinet because at the click of a mouse, the contents can be reshuffled and reorganized. At computer-quick speeds, databases can sort information by any characteristics and filter data into multiple categories.

Helpful Hints

- Before setting up a database, take some time to learn the features of your database software by practicing with established database software.
- Periodically save your database as you enter data. That way, if something happens such as your computer malfunctions or the power goes off, you won't lose all of your work.

Doing a Database Search

When searching for information in a database, use the following search strategies to get the best results. These are the same search methods used for searching internet databases.

- Place the word *and* between two words in your search if you want the database to look for any entries that have both the words. For example, "gold *and* silver" would give you information that mentions both gold and silver.
- Place the word *or* between two words if you want the database to show entries that have at least one of the words. For example "gold *or* silver" would show you information that mentions either gold or silver.
- Place the word *not* between two words if you want the database to look for entries that have the first word but do not have the second word. For example, "gold *not* jewelry" would show you information that mentions gold but does not mention jewelry.

In summary, databases can be used to store large amounts of information about a particular subject. Databases allow biologists, Earth scientists, and physical scientists to search for information quickly and accurately.

Using an Electronic Spreadsheet

Your science fair experiment has produced lots of numbers. How do you keep track of all the data, and how can you easily work out all the calculations needed? You can use a computer program called a spreadsheet to record data that involve numbers. A spreadsheet is an electronic mathematical worksheet.

Type your data in rows and columns, just as they would look in a data table on a sheet of paper. A spreadsheet uses simple math to do data calculations. For example, you could add, subtract, divide, or multiply any of the values in the spreadsheet by another number. You also could set up a series of math steps you want to apply to the data. If you want to add 12 to all the numbers and then multiply all the numbers by 10, the computer does all the calculations for you in the spreadsheet. Below is an example of a spreadsheet that records test car data.

Helpful Hints

- Before you set up the spreadsheet, identify how you want to organize the data. Include any formulas you will need to use.
- Make sure you have entered the correct data into the correct rows and columns.
- You also can display your results in a graph. Pick the style of graph that best represents the data with which you are working.

Figure 19
A spreadsheet allows you to display large amounts of data and do calculations automatically.

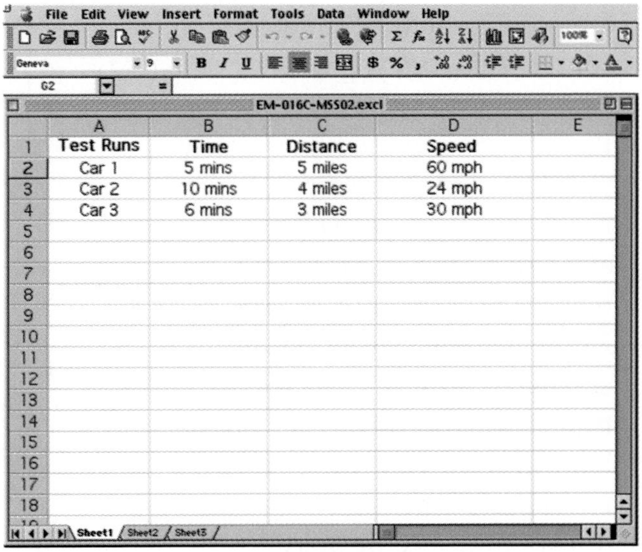

Using a Computerized Card Catalog

When you have a report or paper to research, you probably go to the library. To find the information you need in the library, you might have to use a computerized card catalog. This type of card catalog allows you to search for information by subject, by title, or by author. The computer then will display all the holdings the library has on the subject, title, or author requested.

A library's holdings can include books, magazines, databases, videos, and audio materials. When you have chosen something from this list, the computer will show whether an item is available and where in the library to find it.

Helpful Hints

- Remember that you can use the computer to search by subject, author, or title. If you know a book's author but not the title, you can search for all the books the library has by that author.
- When searching by subject, it's often most helpful to narrow your search by using specific search terms, such as *and, or,* and *not*. If you don't find enough sources this way, you can broaden your search.
- Pay attention to the type of materials found in your search. If you need a book, you can eliminate any videos or other resources that come up in your search.
- Knowing how your library is arranged can save you a lot of time. If you need help, the librarian will show you where certain types of materials are kept and how to find specific holdings.

Using Graphics Software

Are you having trouble finding that exact piece of art you're looking for? Do you have a picture in your mind of what you want but can't seem to find the right graphic to represent your ideas? To solve these problems, you can use graphics software. Graphics software allows you to create and change images and diagrams in almost unlimited ways. Typical uses for graphics software include arranging clip art, changing scanned images, and constructing pictures from scratch. Most graphics software applications work in similar ways. They use the same basic tools and functions. Once you master one graphics application, you can use other graphics applications.

Figure 20
Graphics software can use your data to draw bar graphs.

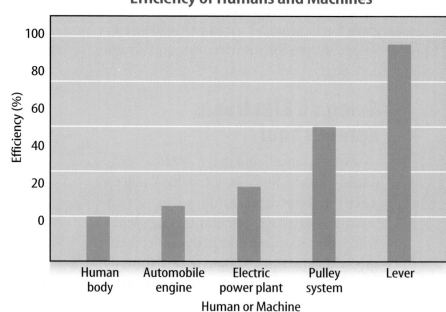

Efficiency of Humans and Machines

Figure 21
Graphics software can use your data to draw circle graphs.

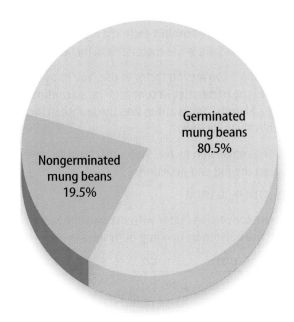

Germinated
mung beans
80.5%

Nongerminated
mung beans
19.5%

Helpful Hints

- As with any method of drawing, the more you practice using the graphics software, the better your results will be.
- Start by using the software to manipulate existing drawings. Once you master this, making your own illustrations will be easier.
- Clip art is available on CD-ROMs and the Internet. With these resources, finding a piece of clip art to suit your purposes is simple.
- As you work on a drawing, save it often.

Developing Multimedia Presentations

It's your turn—you have to present your science report to the entire class. How do you do it? You can use many different sources of information to get the class excited about your presentation. Posters, videos, photographs, sound, computers, and the Internet can help show your ideas.

First, determine what important points you want to make in your presentation. Then, write an outline of what materials and types of media would best illustrate those points. Maybe you could start with an outline on an overhead projector, then show a video, followed by something from the Internet or a slide show accompanied by music or recorded voices. You might choose to use a presentation builder computer application that can combine all these elements into one presentation. Make sure the presentation is well constructed to make the most impact on the audience.

Figure 22
Multimedia presentations use many types of print and electronic materials.

Helpful Hints

- Carefully consider what media will best communicate the point you are trying to make.
- Make sure you know how to use any equipment you will be using in your presentation.
- Practice the presentation several times.
- If possible, set up all of the equipment ahead of time. Make sure everything is working correctly.

Math Skill Handbook

Use this Math Skill Handbook to help solve problems you are given in this text. You might find it useful to review topics in this Math Skill Handbook first.

Converting Units

In science, quantities such as length, mass, and time sometimes are measured using different units. Suppose you want to know how many miles are in 12.7 km?

Conversion factors are used to change from one unit of measure to another. A conversion factor is a ratio that is equal to one. For example, there are 1,000 mL in 1 L, so 1,000 mL equals 1 L, or:

$$1{,}000 \text{ mL} = 1 \text{ L}$$

If both sides are divided by 1 L, this equation becomes:

$$\frac{1{,}000 \text{ mL}}{1 \text{ L}} = 1$$

The **ratio** on the left side of this equation is equal to one and is a conversion factor. You can make another conversion factor by dividing both sides of the top equation by 1,000 mL:

$$1 = \frac{1 \text{ L}}{1{,}000 \text{ mL}}$$

To **convert units,** you multiply by the appropriate conversion factor. For example, how many milliliters are in 1.255 L? To convert 1.255 L to milliliters, multiply 1.255 L by a conversion factor.

Use the **conversion factor** with new units (mL) in the numerator and the old units (L) in the denominator.

$$1.255 \text{ L} \times \frac{1{,}000 \text{ mL}}{1 \text{ L}} = 1{,}255 \text{ mL}$$

The unit L divides in this equation, just as if it were a number.

Example 1 There are 2.54 cm in 1 inch. If a meterstick has a length of 100 cm, how long is the meterstick in inches?

Step 1 Decide which conversion factor to use. You know the length of the meterstick in centimeters, so centimeters are the old units. You want to find the length in inches, so inch is the new unit.

Step 2 Form the conversion factor. Start with the relationship between the old and new units.

$$2.54 \text{ cm} = 1 \text{ inch}$$

Step 3 Form the conversion factor with the old unit (centimeter) on the bottom by dividing both sides by 2.54 cm.

$$1 = \frac{2.54 \text{ cm}}{2.54 \text{ cm}} = \frac{1 \text{ inch}}{2.54 \text{ cm}}$$

Step 4 Multiply the old measurement by the conversion factor.

$$100 \text{ cm} \times \frac{1 \text{ inch}}{2.54 \text{ cm}} = 39.37 \text{ inches}$$

The meter stick is 39.37 inches long.

Example 2 There are 365 days in one year. If a person is 14 years old, what is his or her age in days? (Ignore leap years)

Step 1 Decide which conversion factor to use. You want to convert years to days.

Step 2 Form the conversion factor. Start with the relation between the old and new units.

$$1 \text{ year} = 365 \text{ days}$$

Step 3 Form the conversion factor with the old unit (year) on the bottom by dividing both sides by 1 year.

$$1 = \frac{1 \text{ year}}{1 \text{ year}} = \frac{365 \text{ days}}{1 \text{ year}}$$

Step 4 Multiply the old measurement by the conversion factor:

$$14 \text{ years} \times \frac{365 \text{ days}}{1 \text{ year}} = 5{,}110 \text{ days}$$

The person's age is 5,110 days.

Practice Problem A book has a mass of 2.31 kg. If there are 1,000 g in 1 kg, what is the mass of the book in grams?

Using Fractions

A **fraction** is a number that compares a part to the whole. For example, in the fraction $\frac{2}{3}$, the 2 represents the part and the 3 represents the whole. In the fraction $\frac{2}{3}$, the top number, 2, is called the numerator. The bottom number, 3, is called the denominator.

Sometimes fractions are not written in their simplest form. To determine a fraction's **simplest form,** you must find the greatest common factor (GCF) of the numerator and denominator. The greatest common factor is the largest factor that is common to the numerator and denominator.

For example, because the number 3 divides into 12 and 30 evenly, it is a common factor of 12 and 30. However, because the number 6 is the largest number that evenly divides into 12 and 30, it is the **greatest common factor.**

After you find the greatest common factor, you can write a fraction in its simplest form. Divide both the numerator and the denominator by the greatest common factor. The number that results is the fraction in its **simplest form.**

Example Twelve of the 20 chemicals used in the science lab are in powder form. What fraction of the chemicals used in the lab are in powder form?

Step 1 Write the fraction.

$$\frac{\text{part}}{\text{whole}} = \frac{12}{20}$$

Step 2 To find the GCF of the numerator and denominator, list all of the factors of each number.

Factors of 12: 1, 2, 3, 4, 6, 12 (the numbers that divide evenly into 12)

Factors of 20: 1, 2, 4, 5, 10, 20 (the numbers that divide evenly into 20)

Step 3 List the common factors.

1, 2, 4.

Step 4 Choose the greatest factor in the list of common factors.

The GCF of 12 and 20 is 4.

Step 5 Divide the numerator and denominator by the GCF.

$$\frac{12 \div 4}{20 \div 4} = \frac{3}{5}$$

In the lab, $\frac{3}{5}$ of the chemicals are in powder form.

Practice Problem There are 90 rides at an amusement park. Of those rides, 66 have a height restriction. What fraction of the rides has a height restriction? Write the fraction in simplest form.

Math Skill Handbook

Calculating Ratios

A **ratio** is a comparison of two numbers by division.

Ratios can be written 3 to 5 or 3:5. Ratios also can be written as fractions, such as $\frac{3}{5}$. Ratios, like fractions, can be written in simplest form. Recall that a fraction is in **simplest form** when the greatest common factor (GCF) of the numerator and denominator is 1.

Example A chemical solution contains 40 g of salt and 64 g of baking soda. What is the ratio of salt to baking soda as a fraction in simplest form?

Step 1 Write the ratio as a fraction. $\dfrac{\text{salt}}{\text{baking soda}} = \dfrac{40}{64}$

Step 2 Express the fraction in simplest form. The GCF of 40 and 64 is 8.

$$\frac{40}{64} = \frac{40 \div 8}{64 \div 8} = \frac{5}{8}$$

The ratio of salt to baking soda in the solution is $\frac{5}{8}$.

Practice Problem Two metal rods measure 100 cm and 144 cm in length. What is the ratio of their lengths in simplest fraction form?

Using Decimals

A **decimal** is a fraction with a denominator of 10, 100, 1,000, or another power of 10. For example, 0.854 is the same as the fraction $\frac{854}{1,000}$.

In a decimal, the decimal point separates the ones place and the tenths place. For example, 0.27 means twenty-seven hundredths, or $\frac{27}{100}$, where 27 is the **number of units** out of 100 units. Any fraction can be written as a decimal using division.

Example Write $\frac{5}{8}$ as a decimal.

Step 1 Write a division problem with the numerator, 5, as the dividend and the denominator, 8, as the divisor. Write 5 as 5.000.

Step 2 Solve the problem.

```
        0.625
    8)5.000
      48
      ──
       20
       16
       ──
        40
        40
        ──
         0
```

Therefore, $\frac{5}{8} = 0.625$.

Practice Problem Write $\frac{19}{25}$ as a decimal.

Using Percentages

The word *percent* means "out of one hundred." A **percent** is a ratio that compares a number to 100. Suppose you read that 77 percent of Earth's surface is covered by water. That is the same as reading that the fraction of Earth's surface covered by water is $\frac{77}{100}$. To express a fraction as a percent, first find an equivalent decimal for the fraction. Then, multiply the decimal by 100 and add the percent symbol. For example, $\frac{1}{2} = 1 \div 2 = 0.5$. Then $0.5 = 0.50 = 50\%$.

Example Express $\frac{13}{20}$ as a percent.

Step 1 Find the equivalent decimal for the fraction.

$$
\begin{array}{r}
0.65 \\
20\overline{)13.00} \\
\underline{120} \\
100 \\
\underline{100} \\
0
\end{array}
$$

Step 2 Rewrite the fraction $\frac{13}{20}$ as 0.65.

Step 3 Multiply 0.65 by 100 and add the % sign.

$0.65 \cdot 100 = 65 = 65\%$

So, $\frac{13}{20} = 65\%$.

Practice Problem In one year, 73 of 365 days were rainy in one city. What percent of the days in that city were rainy?

Using Precision and Significant Digits

When you make a **measurement,** the value you record depends on the precision of the measuring instrument. When adding or subtracting numbers with different precision, the answer is rounded to the smallest number of decimal places of any number in the sum or difference. When multiplying or dividing, the answer is rounded to the smallest number of significant figures of any number being multiplied or divided. When counting the number of **significant figures,** all digits are counted except zeros at the end of a number with no decimal such as 2,500, and zeros at the beginning of a decimal such as 0.03020.

Example The lengths 5.28 and 5.2 are measured in meters. Find the sum of these lengths and report the sum using the least precise measurement.

Step 1 Find the sum.

5.28 m	2 digits after the decimal
+ 5.2 m	1 digit after the decimal
10.48 m	

Step 2 Round to one digit after the decimal because the least number of digits after the decimal of the numbers being added is 1.

The sum is 10.5 m.

Practice Problem Multiply the numbers in the example using the rule for multiplying and dividing. Report the answer with the correct number of significant figures.

Math Skill Handbook

An **equation** is a statement that two things are equal. For example, $A = B$ is an equation that states that A is equal to B.

Sometimes one side of the equation will contain a **variable** whose value is not known. In the equation $3x = 12$, the variable is x.

The equation is solved when the variable is replaced with a value that makes both sides of the equation equal to each other. For example, the solution of the equation $3x = 12$ is $x = 4$. If the x is replaced with 4, then the equation becomes $3 \cdot 4 = 12$, or $12 = 12$.

To solve an equation such as $8x = 40$, divide both sides of the equation by the number that multiplies the variable.

$$8x = 40$$
$$\frac{8x}{8} = \frac{40}{8}$$
$$x = 5$$

You can check your answer by replacing the variable with your solution and seeing if both sides of the equation are the same.

$$8x = 8 \cdot 5 = 40$$

The left and right sides of the equation are the same, so $x = 5$ is the solution.

Sometimes an equation is written in this way: $a = bc$. This also is called a **formula.** The letters can be replaced by numbers, but the numbers must still make both sides of the equation the same.

Example 1 Solve the equation $10x = 35$.

Step 1 Find the solution by dividing each side of the equation by 10.

$$10x = 35 \qquad \frac{10x}{10} = \frac{35}{10} \qquad x = 3.5$$

Step 2 Check the solution.

$$10x = 35 \qquad 10 \times 3.5 = 35 \qquad 35 = 35$$

Both sides of the equation are equal, so $x = 3.5$ is the solution to the equation.

Example 2 In the formula $a = bc$, find the value of c if $a = 20$ and $b = 2$.

Step 1 Rearrange the formula so the unknown value is by itself on one side of the equation by dividing both sides by b.

$$a = bc$$
$$\frac{a}{b} = \frac{bc}{b}$$
$$\frac{a}{b} = c$$

Step 2 Replace the variables a and b with the values that are given.

$$\frac{a}{b} = c$$
$$\frac{20}{2} = c$$
$$10 = c$$

Step 3 Check the solution.

$$a = bc$$
$$20 = 2 \times 10$$
$$20 = 20$$

Both sides of the equation are equal, so $c = 10$ is the solution when $a = 20$ and $b = 2$.

Practice Problem In the formula $h = gd$, find the value of d if $g = 12.3$ and $h = 17.4$.

Using Proportions

A **proportion** is an equation that shows that two ratios are equivalent. The ratios $\frac{2}{4}$ and $\frac{5}{10}$ are equivalent, so they can be written as $\frac{2}{4} = \frac{5}{10}$. This equation is an example of a proportion.

When two ratios form a proportion, the **cross products** are equal. To find the cross products in the proportion $\frac{2}{4} = \frac{5}{10}$, multiply the 2 and the 10, and the 4 and the 5. Therefore $2 \cdot 10 = 4 \cdot 5$, or $20 = 20$.

Because you know that both proportions are equal, you can use cross products to find a missing term in a proportion. This is known as **solving the proportion.** Solving a proportion is similar to solving an equation.

Example The heights of a tree and a pole are proportional to the lengths of their shadows. The tree casts a shadow of 24 m at the same time that a 6-m pole casts a shadow of 4 m. What is the height of the tree?

Step 1 Write a proportion.

$$\frac{\text{height of tree}}{\text{height of pole}} = \frac{\text{length of tree's shadow}}{\text{length of pole's shadow}}$$

Step 2 Substitute the known values into the proportion. Let h represent the unknown value, the height of the tree.

$$\frac{h}{6} = \frac{24}{4}$$

Step 3 Find the cross products.

$$h \cdot 4 = 6 \cdot 24$$

Step 4 Simplify the equation.

$$4h = 144$$

Step 5 Divide each side by 4.

$$\frac{4h}{4} = \frac{144}{4}$$

$$h = 36$$

The height of the tree is 36 m.

Practice Problem The ratios of the weights of two objects on the Moon and on Earth are in proportion. A rock weighing 3 N on the Moon weighs 18 N on Earth. How much would a rock that weighs 5 N on the Moon weigh on Earth?

Math Skill Handbook

Statistics is the branch of mathematics that deals with collecting, analyzing, and presenting data. In statistics, there are three common ways to summarize the data with a single number—the mean, the median, and the mode.

The **mean** of a set of data is the arithmetic average. It is found by adding the numbers in the data set and dividing by the number of items in the set.

The **median** is the middle number in a set of data when the data are arranged in numerical order. If there were an even number of data points, the median would be the mean of the two middle numbers.

The **mode** of a set of data is the number or item that appears most often.

Another number that often is used to describe a set of data is the range. The **range** is the difference between the largest number and the smallest number in a set of data.

A **frequency table** shows how many times each piece of data occurs, usually in a survey. The frequency table below shows the results of a student survey on favorite color.

Color	Tally	Frequency
red	||||	4
blue	卌	5
black	||	2
green	|||	3
purple	卌 ||	7
yellow	卌 |	6

Based on the frequency table data, which color is the favorite?

Example The speeds (in m/s) for a race car during five different time trials are 39, 37, 44, 36, and 44.

To find the mean:
Step 1 Find the sum of the numbers.

$$39 + 37 + 44 + 36 + 44 = 200$$

Step 2 Divide the sum by the number of items, which is 5.

$$200 \div 5 = 40$$

The mean measure is 40 m/s.

To find the median:
Step 1 Arrange the measures from least to greatest.

$$36, \ 37, \ \underline{39}, \ 44, \ 44$$

Step 2 Determine the middle measure.

The median measure is 39 m/s.

To find the mode:
Step 1 Group the numbers that are the same together.

$$44, 44, 36, 37, 39$$

Step 2 Determine the number that occurs most in the set.

$$\underline{44, 44}, 36, 37, 39$$

The mode measure is 44 m/s.

To find the range:
Step 1 Arrange the measures from largest to smallest.

$$44, 44, 39, 37, 36$$

Step 2 Determine the largest and smallest measures in the set.

$$\underline{44}, 44, 39, 37, \underline{36}$$

Step 3 Find the difference between the largest and smallest measures.

$$44 - 36 = 8$$

The range is 8 m/s.

Practice Problem Find the mean, median, mode, and range for the data set 8, 4, 12, 8, 11, 14, 16.

Safety in the Science Classroom

1. Always obtain your teacher's permission to begin an investigation.

2. Study the procedure. If you have questions, ask your teacher. Be sure you understand any safety symbols shown on the page.

3. Use the safety equipment provided for you. Goggles and a safety apron should be worn during most investigations.

4. Always slant test tubes away from yourself and others when heating them or adding substances to them.

5. Never eat or drink in the lab, and never use lab glassware as food or drink containers. Never inhale chemicals. Do not taste any substances or draw any material into a tube with your mouth.

6. Report any spill, accident, or injury, no matter how small, immediately to your teacher, then follow his or her instructions.

7. Know the location and proper use of the fire extinguisher, safety shower, fire blanket, first aid kit, and fire alarm.

8. Keep all materials away from open flames. Tie back long hair and tie down loose clothing.

9. If your clothing should catch fire, smother it with the fire blanket, or get under a safety shower. NEVER RUN.

10. If a fire should occur, turn off the gas then leave the room according to established procedures.

Follow these procedures as you clean up your work area

1. Turn off the water and gas. Disconnect electrical devices.

2. Clean all pieces of equipment and return all materials to their proper places.

3. Dispose of chemicals and other materials as directed by your teacher. Place broken glass and solid substances in the proper containers. Make sure never to discard materials in the sink.

4. Clean your work area. Wash your hands thoroughly after working in the laboratory.

First Aid	
Injury	Safe Response ALWAYS NOTIFY YOUR TEACHER IMMEDIATELY
Burns	Apply cold water.
Cuts and Bruises	Stop any bleeding by applying direct pressure. Cover cuts with a clean dressing. Apply ice packs or cold compresses to bruises.
Fainting	Leave the person lying down. Loosen any tight clothing and keep crowds away.
Foreign Matter in Eye	Flush with plenty of water. Use eyewash bottle or fountain.
Poisoning	Note the suspected poisoning agent.
Any Spills on Skin	Flush with large amounts of water or use safety shower.

PERIODIC TABLE OF THE ELEMENTS

Columns of elements are called groups. Elements in the same group have similar chemical properties.

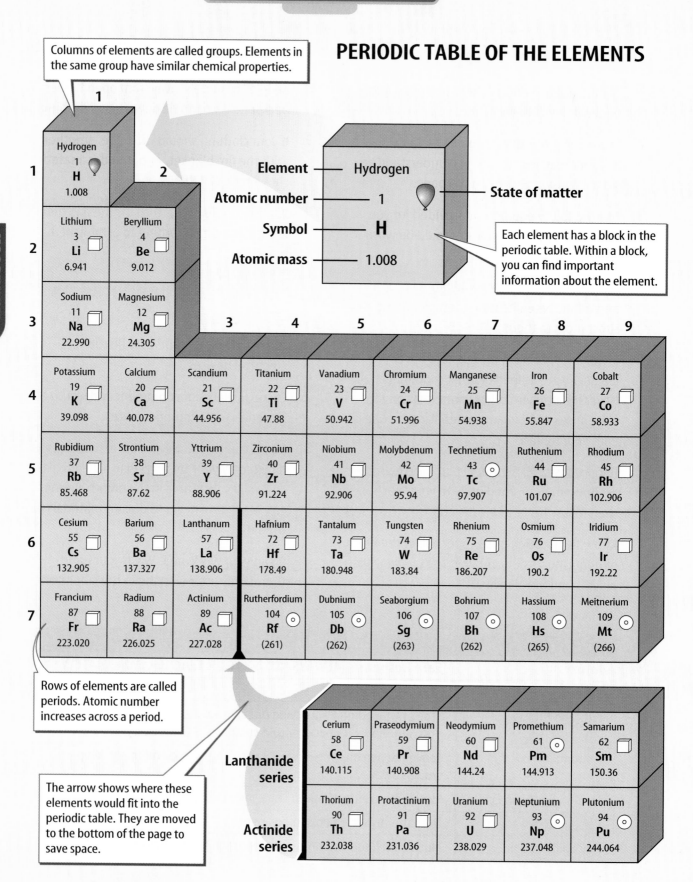

Element —— Hydrogen

Atomic number —— 1

Symbol —— **H**

Atomic mass —— 1.008

State of matter

Each element has a block in the periodic table. Within a block, you can find important information about the element.

1

1 Hydrogen
1
H
1.008

2 Lithium
3
Li
6.941

Beryllium
4
Be
9.012

3 Sodium
11
Na
22.990

Magnesium
12
Mg
24.305

| | **3** | **4** | **5** | **6** | **7** | **8** | **9** |

4 Potassium
19
K
39.098

Calcium
20
Ca
40.078

Scandium
21
Sc
44.956

Titanium
22
Ti
47.88

Vanadium
23
V
50.942

Chromium
24
Cr
51.996

Manganese
25
Mn
54.938

Iron
26
Fe
55.847

Cobalt
27
Co
58.933

5 Rubidium
37
Rb
85.468

Strontium
38
Sr
87.62

Yttrium
39
Y
88.906

Zirconium
40
Zr
91.224

Niobium
41
Nb
92.906

Molybdenum
42
Mo
95.94

Technetium
43
Tc
97.907

Ruthenium
44
Ru
101.07

Rhodium
45
Rh
102.906

6 Cesium
55
Cs
132.905

Barium
56
Ba
137.327

Lanthanum
57
La
138.906

Hafnium
72
Hf
178.49

Tantalum
73
Ta
180.948

Tungsten
74
W
183.84

Rhenium
75
Re
186.207

Osmium
76
Os
190.2

Iridium
77
Ir
192.22

7 Francium
87
Fr
223.020

Radium
88
Ra
226.025

Actinium
89
Ac
227.028

Rutherfordium
104
Rf
(261)

Dubnium
105
Db
(262)

Seaborgium
106
Sg
(263)

Bohrium
107
Bh
(262)

Hassium
108
Hs
(265)

Meitnerium
109
Mt
(266)

Rows of elements are called periods. Atomic number increases across a period.

The arrow shows where these elements would fit into the periodic table. They are moved to the bottom of the page to save space.

Lanthanide series

Cerium
58
Ce
140.115

Praseodymium
59
Pr
140.908

Neodymium
60
Nd
144.24

Promethium
61
Pm
144.913

Samarium
62
Sm
150.36

Actinide series

Thorium
90
Th
232.038

Protactinium
91
Pa
231.036

Uranium
92
U
238.029

Neptunium
93
Np
237.048

Plutonium
94
Pu
244.064

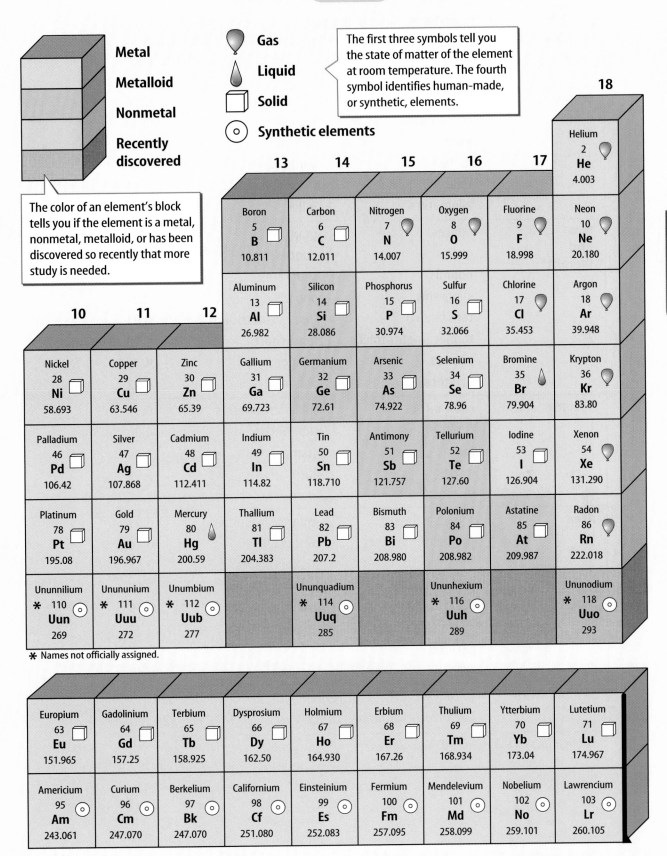

Metal

Metalloid

Nonmetal

Recently discovered

The color of an element's block tells you if the element is a metal, nonmetal, metalloid, or has been discovered so recently that more study is needed.

Gas

Liquid

Solid

Synthetic elements

The first three symbols tell you the state of matter of the element at room temperature. The fourth symbol identifies human-made, or synthetic, elements.

			13	14	15	16	17	18
								Helium 2 He 4.003
			Boron 5 B 10.811	Carbon 6 C 12.011	Nitrogen 7 N 14.007	Oxygen 8 O 15.999	Fluorine 9 F 18.998	Neon 10 Ne 20.180
10	11	12	Aluminum 13 Al 26.982	Silicon 14 Si 28.086	Phosphorus 15 P 30.974	Sulfur 16 S 32.066	Chlorine 17 Cl 35.453	Argon 18 Ar 39.948
Nickel 28 Ni 58.693	Copper 29 Cu 63.546	Zinc 30 Zn 65.39	Gallium 31 Ga 69.723	Germanium 32 Ge 72.61	Arsenic 33 As 74.922	Selenium 34 Se 78.96	Bromine 35 Br 79.904	Krypton 36 Kr 83.80
Palladium 46 Pd 106.42	Silver 47 Ag 107.868	Cadmium 48 Cd 112.411	Indium 49 In 114.82	Tin 50 Sn 118.710	Antimony 51 Sb 121.757	Tellurium 52 Te 127.60	Iodine 53 I 126.904	Xenon 54 Xe 131.290
Platinum 78 Pt 195.08	Gold 79 Au 196.967	Mercury 80 Hg 200.59	Thallium 81 Tl 204.383	Lead 82 Pb 207.2	Bismuth 83 Bi 208.980	Polonium 84 Po 208.982	Astatine 85 At 209.987	Radon 86 Rn 222.018
Ununnilium * 110 Uun 269	Unununium * 111 Uuu 272	Unumbium * 112 Uub 277		Ununquadium * 114 Uuq 285		Ununhexium * 116 Uuh 289		Ununodium * 118 Uuo 293

* Names not officially assigned.

Europium 63 Eu 151.965	Gadolinium 64 Gd 157.25	Terbium 65 Tb 158.925	Dysprosium 66 Dy 162.50	Holmium 67 Ho 164.930	Erbium 68 Er 167.26	Thulium 69 Tm 168.934	Ytterbium 70 Yb 173.04	Lutetium 71 Lu 174.967
Americium 95 Am 243.061	Curium 96 Cm 247.070	Berkelium 97 Bk 247.070	Californium 98 Cf 251.080	Einsteinium 99 Es 252.083	Fermium 100 Fm 257.095	Mendelevium 101 Md 258.099	Nobelium 102 No 259.101	Lawrencium 103 Lr 260.105

Reference Handbook

REFERENCE HANDBOOK C

SI—Metric/English, English/Metric Conversions

	When you want to convert:	To:	Multiply by:
Length	inches	centimeters	2.54
	centimeters	inches	0.39
	yards	meters	0.91
	meters	yards	1.09
	miles	kilometers	1.61
	kilometers	miles	0.62
Mass and Weight*	ounces	grams	28.35
	grams	ounces	0.04
	pounds	kilograms	0.45
	kilograms	pounds	2.2
	tons (short)	tonnes (metric tons)	0.91
	tonnes (metric tons)	tons (short)	1.10
	pounds	newtons	4.45
	newtons	pounds	0.22
Volume	cubic inches	cubic centimeters	16.39
	cubic centimeters	cubic inches	0.06
	liters	quarts	1.06
	quarts	liters	0.95
	gallons	liters	3.78
Area	square inches	square centimeters	6.45
	square centimeters	square inches	0.16
	square yards	square meters	0.83
	square meters	square yards	1.19
	square miles	square kilometers	2.59
	square kilometers	square miles	0.39
	hectares	acres	2.47
	acres	hectares	0.40
Temperature	To convert °Celsius to °Fahrenheit	$°C \times 9/5 + 32$	
	To convert °Fahrenheit to °Celsius	$5/9 \, (°F - 32)$	

*Weight is measured in standard Earth gravity.

REFERENCE Handbook D

SAFETY SYMBOLS

SAFETY SYMBOLS	HAZARD	EXAMPLES	PRECAUTION	REMEDY
DISPOSAL	Special disposal procedures need to be followed.	certain chemicals, living organisms	Do not dispose of these materials in the sink or trash can.	Dispose of wastes as directed by your teacher.
BIOLOGICAL	Organisms or other biological materials that might be harmful to humans	bacteria, fungi, blood, unpreserved tissues, plant materials	Avoid skin contact with these materials. Wear mask or gloves.	Notify your teacher if you suspect contact with material. Wash hands thoroughly.
EXTREME TEMPERATURE	Objects that can burn skin by being too cold or too hot	boiling liquids, hot plates, dry ice, liquid nitrogen	Use proper protection when handling.	Go to your teacher for first aid.
SHARP OBJECT	Use of tools or glassware that can easily puncture or slice skin	razor blades, pins, scalpels, pointed tools, dissecting probes, broken glass	Practice common-sense behavior and follow guidelines for use of the tool.	Go to your teacher for first aid.
FUME	Possible danger to respiratory tract from fumes	ammonia, acetone, nail polish remover, heated sulfur, moth balls	Make sure there is good ventilation. Never smell fumes directly. Wear a mask.	Leave foul area and notify your teacher immediately.
ELECTRICAL	Possible danger from electrical shock or burn	improper grounding, liquid spills, short circuits, exposed wires	Double-check setup with teacher. Check condition of wires and apparatus.	Do not attempt to fix electrical problems. Notify your teacher immediately.
IRRITANT	Substances that can irritate the skin or mucus membranes of the respiratory tract	pollen, moth balls, steel wool, fiber glass, potassium permanganate	Wear dust mask and gloves. Practice extra care when handling these materials.	Go to your teacher for first aid.
CHEMICAL	Chemicals that can react with and destroy tissue and other materials	bleaches such as hydrogen peroxide; acids such as sulfuric acid, hydrochloric acid; bases such as ammonia, sodium hydroxide	Wear goggles, gloves, and an apron.	Immediately flush the affected area with water and notify your teacher.
TOXIC	Substance may be poisonous if touched, inhaled, or swallowed	mercury, many metal compounds, iodine, poinsettia plant parts	Follow your teacher's instructions.	Always wash hands thoroughly after use. Go to your teacher for first aid.
OPEN FLAME	Open flame may ignite flammable chemicals, loose clothing, or hair	alcohol, kerosene, potassium permanganate, hair, clothing	Tie back hair. Avoid wearing loose clothing. Avoid open flames when using flammable chemicals. Be aware of locations of fire safety equipment.	Notify your teacher immediately. Use fire safety equipment if applicable.

Eye Safety
Proper eye protection should be worn at all times by anyone performing or observing science activities.

Clothing Protection
This symbol appears when substances could stain or burn clothing.

Animal Safety
This symbol appears when safety of animals and students must be ensured.

Radioactivity
This symbol appears when radioactive materials are used.

English Glossary

This glossary defines each key term that appears in bold type in the text. It also shows the chapter, section, and page number where you can find the word used.

A

amplitude: distance a wave rises above or falls below its normal level, which is related to the energy that the wave carries; in a transverse wave, is one half the distance be-tween a crest and a trough. (Chap. 1, Sec. 2, p. 13)

C

carrier wave: particular transmission frequency assigned to a radio station. (Chap. 3, Sec. 3, p. 82)

compressional wave: a type of mechanical wave in which matter in the medium moves forward and backward in the same direction the wave travels. (Chap. 1, Sec. 1, p. 11)

concave lens: lens that is thicker at its edges than in the middle and causes light rays traveling parallel to the optical axis to diverge. (Chap. 4, Sec. 3, p. 111)

convex lens: converging lens that is thicker in the middle than at its edges. (Chap. 4, Sec. 3, p. 110)

D

diffraction: bending of waves around a barrier. (Chap. 1, Sec. 3, p. 22)

Doppler effect: change in the frequency or pitch of a sound that occurs when the sound source and the listener are in motion relative to each other. (Chap. 2, Sec. 1, p. 42)

E

eardrum: membrane stretching across the ear canal that vibrates when sound waves reach the middle ear. (Chap. 2, Sec. 2, p. 54)

echo: a reflected sound wave. (Chap. 2, Sec. 1, p. 41)

electromagnetic spectrum: range of electromagnetic waves, including radio waves, visible light, and X rays, with different frequencies and wavelengths. (Chap. 3, Sec. 2, p. 71)

electromagnetic waves: transverse waves that can travel through matter or space, are produced by the motion of electrically charged particles, and include X rays, ultraviolet waves, and visible light. (Chap. 1, Sec. 1, p. 12)

F

focal length: distance along the optical axis from the center of a concave mirror to the focal point. (Chap. 4, Sec. 2, p. 104)

focal point: single point on the optical axis of a concave mirror where re-flected light rays pass through. (Chap. 4, Sec. 2, p. 104)

frequency: number of wavelengths that pass a given point in one second, measured in hertz (Hz). (Chap. 1, Sec. 2, p. 15)

fundamental frequency: lowest natural frequency that is produced by a vibrating string or vibrating column of air. (Chap. 2, Sec. 2, p. 49)

G

gamma ray: highest-frequency, most penetrating electromagnetic wave. (Chap. 3, Sec. 2, p. 76)

Global Positioning System (GPS): uses satellites, ground-based stations, and portable units with receivers to locate objects on Earth. (Chap. 3, Sec. 3, p. 85)

I

infrared wave: electromagnetic wave that is sensed as heat and is emitted by almost every object. (Chap. 3, Sec. 2, p. 73)

interference: ability of two or more waves to combine and form a new wave when they overlap. (Chap. 1, Sec. 3, p. 24)

L

law of reflection: states that the angle of incidence is equal to the angle of reflection. (Chap. 4, Sec. 2, p. 101)

lens: transparent object that has at least one curved surface that causes light to bend. (Chap. 4, Sec. 3, p. 109)

light ray: narrow beam of light traveling in a straight line. (Chap. 4, Sec. 1, p. 96)

loudness: the human perception of how much energy a sound wave carries. (Chap. 2, Sec. 1, p. 38)

M

mechanical wave: a type of wave that can travel only through matter. (Chap. 1, Sec. 1, p. 9)

medium: material through which a wave can travel. (Chap. 4, Sec. 1, p. 97)

N

natural frequency: frequency at which a musical instrument or other object vibrates when it is struck or disturbed; relative to its size, shape, and the material it is made from. (Chap. 2, Sec. 2, p. 47)

O

overtones: multiples of the fundamental frequency. (Chap. 2, Sec. 2, p. 49)

P

pitch: how high or low a sound is. (Chap. 2, Sec. 1, p. 40)

R

radiant energy: energy carried by an electromagnetic wave. (Chap. 3, Sec. 1, p. 70)

radio waves: lowest-frequency electromagnetic waves that carry the least amount of energy and are used in most forms of tele-com-munications technology—such as TVs, telephones, and radios. (Chap. 3, Sec. 2, p. 72)

reflecting telescope: uses a concave mirror to gather light from distant objects. (Chap. 4, Sec. 4, p. 115)

reflection: occurs when a wave strikes an object or surface and bounces off. (Chap. 1, Sec. 3, p. 20) (Chap. 4, Sec. 1, p. 97)

refracting telescope: uses two convex lenses to gather light and form an image of a distant object. (Chap. 4, Sec. 4, p. 114)

refraction: bending of a wave as it moves from one medium into another medium; due to a change in speed. (Chap. 1, Sec. 3, p. 21) (Chap. 4, Sec.3, p. 109)

resonance: sound amplification that occurs when an object is vibrated at its natural frequency by absorbing energy from a sound wave or other object vibrating at this frequency. (Chap. 2, Sec. 2, p. 48)

reverberation: repeated echoes of sounds. (Chap. 2, Sec. 2, p. 53)

T

transverse wave: a type of mechanical wave in which the wave energy causes matter in the medium to move up and down or back and forth at right angles to the direction the wave travels. (Chap. 1, Sec. 1, p. 10)

U

ultraviolet radiation (UV): electromagnetic waves with higher frequencies and shorter wavelengths than visible light. (Chap. 3, Sec. 2, p. 75)

V

visible light: electromagnetic waves with wavelengths between 0.4 and 0.7 millionths of a meter that can be seen with your eyes. (Chap. 3, Sec. 2, p. 74)

W

wave: rhythmic disturbance that carries energy but not matter. (Chap. 1, Sec. 1, p. 8)

wavelength: in transverse waves, the distance between the tops of two adjacent crests or the bottoms of two adjacent troughs; in compressional waves, the distance from the centers of adjacent rarefactions. (Chap. 1, Sec. 2, p. 14)

X

X ray: high-energy electromagnetic wave that is highly penetrating and can be used for medical diagnosis. (Chap. 3, Sec. 2, p. 76)

English Glossary

Este glosario define cada término clave que aparece en negrillas en el texto. También muestra el capítulo, la sección y el número de página en donde se usa dicho término.

A

amplitude / amplitud: distancia a la cual una onda sube o baja de su nivel normal, la cual se relaciona con la energía que transporta la onda; en una onda transversal, es la mitad de la distancia entre una cresta y un seno. (Cap. 1, Sec. 2, pág. 13)

C

carrier wave / onda portadora: frecuencia de transmisión particular asignada a una estación radial. (Cap. 3, Sec. 3, pág. 82)

compressional wave / onda de compresión: tipo de onda mecánica en la cual la materia del medio oscila en la misma dirección en que viaja la onda. (Cap. 1, Sec. 1, pág. 11)

concave lens / lente cóncavo: lente que es más gruesa en los bordes que en el medio y que desvía los rayos luminosos que viajan paralelos al eje óptico. (Cap. 4, Sec. 3, pág. 111)

convex lens / lente convexo: lente convergente que es más gruesa en el medio que en los bordes. (Cap. 4, Sec. 3, pág. 110)

D

diffraction / difracción: desviación de las ondas alrededor de un obstáculo. (Cap. 1, Sec. 3, pág. 22)

Doppler effect / efecto Doppler: cambio en la frecuencia o el tono de un sonido, el cual ocurre cuando la fuente sonora y el oyente están en movimiento relativo uno del otro. (Cap. 2, Sec. 1, pág. 42)

E

eardrum / tímpano: membrana que se extiende a través del canal auditivo y la cual vibra cuando las ondas sonoras llegan al oído medio. (Cap. 2, Sec. 2, pág. 54)

echo / eco: onda sonora reflejada. (Cap. 2, Sec. 1, pág. 41)

electromagnetic spectrum / espectro electromagnético: rango de ondas electromagnéticas, que incluyen las ondas radiales, la luz visible y los rayos X, las cuales poseen distintas frecuencias y longitudes de onda. (Cap. 3, Sec. 2, pág. 71)

electromagnetic waves / ondas electromagnéticas: ondas transversales que pueden viajar a través de la materia o el espacio, se producen debido al movimiento de partículas cargadas eléctricamente e incluyen los rayos X, las ondas ultravioletas y la luz visible. (Cap. 1, Sec. 1, pág. 12)

F

focal length / longitud focal: distancia a lo largo del eje óptico desde el centro de un espejo cóncavo al punto focal. (Cap. 4, Sec. 2, pág. 104)

focal point / punto focal: punto único en el eje óptico de un espejo cóncavo a través del cual pasan los rayos luminosos reflejados. (Cap. 4, Sec. 2, pág. 104)

frequency / frecuencia: número de longitudes de onda que pasan por un punto dado en un segundo; se miden en hertz (Hz). (Cap. 1, Sec. 2, pág. 15)

fundamental frequency / frecuencia fundamental: frecuencia natural más baja que produce una cuerda o una columna de aire que vibra. (Cap. 2, Sec. 2, pág. 49)

G

gamma ray / rayo gama: la onda electromagnética más penetrante y de alta frecuencia. (Cap. 3, Sec. 2, pág. 76)

Global Positioning System / Sistema de Posición Global: usa satélites, estaciones terrestres y equipo portátil con receptores para ubicar objetos sobre la Tierra. (Cap. 3, Sec. 3, pág. 85)

I

infrared wave / onda infrarroja: onda electromagnética que se siente como calor y la cual emiten casi todos los objetos. (Cap. 3, Sec. 2, pág. 73)

interference / interferencia: capacidad de dos o más ondas de combinarse y formar una nueva onda cuando se traslapan. (Cap. 1, Sec. 3, pág. 24)

L

law of reflection / ley de la reflexión: establece que el ángulo de incidencia es igual al ángulo de reflexión. (Cap. 4, Sec. 2, pág. 101)

lens / lente: objeto transparente que tiene por lo menos una superficie que hace que la luz se doble. (Cap. 4, Sec. 3, pág. 109)

light ray / rayo luminoso: rayo angosto de luz que viaja en línea recta. (Cap. 4, Sec. 1, pág. 96)

loudness / volumen: el grado de fortaleza de un sonido; se mide según la amplitud (grado de compresión) de una onda sonora y se puede describir según la escala de decibeles. (Cap. 2, Sec. 1, pág. 38)

M

mechanical wave / onda mecánica: tipo de onda que sólo puede viajar a través de la materia. (Cap. 1, Sec. 1, pág. 9)

medium / medio: material a través del cual puede viajar una onda. (Cap. 4, Sec. 1, pág. 97)

N

natural frequency / frecuencia natural: frecuencia a la cual vibra un instrumento musical u otro objeto cuando se puntea o se perturba, con relación a su tamaño, forma y el material del cual está hecho. (Cap. 2, Sec. 2, pág. 47)

O

overtones / sobretonos: múltiplos de la frecuencia fundamental. (Cap. 2, Sec. 2, pág. 49)

P

pitch / tono: el grado de agudeza o gravedad de un sonido. (Cap. 2, Sec. 1, pág. 40)

R

radiant energy / energía radiante: energía que transportan las ondas electromagnéticas. (Cap. 3, Sec. 1, pág. 70)

radio waves / ondas radiales: ondas electromagnéticas de la más baja frecuencia que transportan la menor cantidad de energía y las cuales se utilizan en casi todas las formas de telecomunicaciones, por ejemplo, los televisores, los teléfonos y los radios. (Cap. 3, Sec. 2, pág. 72)

reflecting telescope / telescopio reflector: usa un espejo cóncavo para recoger la luz de objetos distantes. (Cap. 4, Sec. 4, pág. 115)

reflection / reflexión: ocurre cuando una onda choca contra un cuerpo o una superficie y rebota. (Cap. 1, Sec. 3, pág. 20; Cap. 4, Sec. 1, pág. 97)

refracting telescope / telescopio refractor: usa dos lentes convexas para recoger la luz y formar una imagen de un objeto distante. (Cap. 4, Sec. 4, pág. 114)

refraction / refracción: describe cómo se doblan las ondas luminosas cuando pasan de un medio a otro medio, debido a un cambio de velocidad. (Cap. 1, Sec. 3, pág. 21; Cap. 4, Sec. 3, pág. 109)

resonance / resonancia: amplificación sonora que ocurre cuando un objeto vibra a su frecuencia natural al absorber energía de una onda sonora u otro objeto que vibra a esa misma frecuencia. (Cap. 2, Sec. 2, pág. 48)

reverberation / reverberación: ecos de sonidos repetidos. (Cap. 2, Sec. 2, pág. 53)

T

transverse wave / onda transversal: tipo de onda mecánica en la cual la energía de la onda hace que la materia del medio suba o baje u oscile formando ángulos rectos con la dirección en que viaja la onda. (Cap. 1, Sec. 1, pág. 10)

U

ultraviolet radiation(UV) / radiación ultravioleta(RU): ondas electromagnéticas con frecuencias más altas y longitudes de onda más cortas que la luz visible. (Cap. 3, Sec. 2, pág. 75)

V

visible light / luz visible: ondas electro-magnéticas con longitudes de onda entre 0.4 y 0.7 millonésimas de metro y las cuales se pueden ver a simple vista. (Cap. 3, Sec. 2, pág. 74)

W

wave / onda: perturbación rítmica que transporta energía pero no materia. (Cap. 1, Sec. 1, pág. 8)

wavelength / longitud de onda: en las ondas transversales, es la distancia entre la parte superior de dos crestas adya-centes o la parte inferior de dos senos adyacentes; en las ondas de compresión, es la distancia desde los centros de rare-facciones adyacentes. (Cap. 1, Sec. 2, pág. 14)

X

X ray / rayo X: onda electromagnética de alta frecuencia que es muy penetrante y la cual se usa en el diagnóstico médico. (Cap. 3, Sec. 2, pág. 76)

The index for *Waves, Sound, and Light* will help you locate major topics in the book quickly and easily. Each entry in the index is followed by the number of the pages on which the entry is discussed. A page number given in boldfaced type indicates the page on which that entry is defined. A page number given in italic type indicates a page on which the entry is used in an illustration or photograph. The abbreviation *act.* indicates a page on which the entry is used in an activity.

Index

Art Credits

Glencoe would like to acknowledge the artists and agencies who participated in illustrating this program: Absolute Science Illustration; Andrew Evansen; Argosy; Articulate Graphics; Craig Attebery represented by Frank & Jeff Lavaty; CHK America; Gagliano Graphics; Pedro Julio Gonzalez represented by Melissa Turk & The Artist Network; Robert Hynes represented by Mendola Ltd.; Morgan Cain & Associates; JTH Illustration; Laurie O'Keefe; Matthew Pippin represented by Beranbaum Artist's Representative; Precision Graphics; Publisher's Art; Rolin Graphics, Inc.; Wendy Smith represented by Melissa Turk & The Artist Network; Kevin Torline represented by Berendsen and Associates, Inc.; WILDlife ART; Phil Wilson represented by Cliff Knecht Artist Representative; Zoo Botanica.

Photo Credits

Abbreviation key: AA=Animals Animals; AH=Aaron Haupt; AMP=Amanita Pictures; BC=Bruce Coleman, Inc.; CB=CORBIS; DM=Doug Martin; DRK=DRK Photo; ES=Earth Scenes; FP=Fundamental Photographs; GH=Grant Heilman Photography; IC=Icon Images; KS=KS Studios; LA=Liaison Agency; MB=Mark Burnett; MM=Matt Meadows; PE=PhotoEdit; PD=PhotoDisc; PQ=PictureQuest; PR=Photo Researchers; SB=Stock Boston; TSA=Tom Stack & Associates; TSM=The Stock Market; VU=Visuals Unlimited.

Cover PD; V Cary Wolinsky/SB/PQ; **vi** (t)Charles O'Rear/CB, (b)Timothy Fuller; **1** Douglas Peebles/CB; **2** Bettmann/CB; **3** (t)Schnectady Museum/Hall of Electrical History Foundation/CB, (bl)U.S. Department of the Interior, National Park Service, Edison National Historic Site, (br)CB; **4** Schnectady Museum/Hall of Electrical History Foundation/CB; **6** Jerome Wexler/PR; **6-7** Douglas Peebles/CB; **7** Spencer Grant/PE; **8** (l)file photo, (r)David Young-Wolff/PE; **9** David Young-Wolff/PE; **10** Mark Thayer; **13** Steven Starr/SB; **18** Ken Frick; **19** MB; **21** Ernst Haas/Stone; **22** Peter Beattie/LA; **24** (t)D. Boone/CB, (b)Stephen R. Wagner; **25** Seth Resnick/SB; **26** (t)Reuters NewMedia/CB, (b)Timothy Fuller; **28** (t)John Evans, (b)SuperStock; **29** Roger Ressmeyer/CB; **30** Mark Thayer; **34** Paul Silverman/FP; **34-35** Roger Ressmeyer/CB; **35** Timothy Fuller; **39** (t)Joe Towers/TSM, (c)Bob Daemmrich/SB/PQ, (b)Jean-Paul Thomas/Jacana Scientific Control/PR; **41** Stephen Dalton/PR; **42** NOAA; **43** Slim Films; **45** Spencer Grant/PE; **46** Timothy Fuller; **48** Mark Thayer; **50** Dilip Mehta/Contact Press Images/PQ; **51** (tl)CB, (tr)Paul Seheult/Eye Ubiquitous/CB, (b)IC; **52** William Whitehurst/TSM; **53** SuperStock; **54** (t)Geostock/PD, (b)SuperStock; **55** Fred E. Hossler/VU; **56** (t)Ryan McVay/PD, (b)Oliver Benn/Stone; **58** Douglas Whyte/TSM; **59** (t)Steve Labadessa/Time, (c)courtesy 3M, (b)Bernard Roussel/The Image Bank; **60** (t)Edmond Van Hoorick/PD, (c)Will McIntyre/PR, (b)Kim Steele/PD; **61** (tl)The Photo Works/PR, (tr)PD, (cl)Artville, (cr)PD, (b)Gary Braasch/Stone; **62** PhotoSpin/Artville/PQ; **63** C. Squared Studios/PD; **64** Stephanie Maze/CB; **64-65** Roger Ressmeyer/CB; **65** IC; **66** (l)Bob Abraham/TSM, (r)Jeff Greenberg/VU; **67** (l)David Young-Wolff/PE, (r)NRSC, Ltd./Science Photo Library/PR; **68** (t)Grantpix/PR, (b)Richard Megna/FP; **70** Luke Dodd/Science Photo Library/PR; **72** (t)MM, (b)Jean Miele/TSM; **74** (t)Gregory G.

Dimijian/PR, (b)Charlie Westerman/LA; **75** AH; **77** (l)MM, (r)Bob Daemmrich/The Image Works; **78** (t cl)NASA, (cr)Max Planck Institute for Radio Astronomy/Science Photo Library/PR, (b)ESA/Science Photo Library/PR; **79** (l)NASA/Science Photo Library/PR, (c)Harvard-Smithsonian Center for Astrophysics, (r)ESA; **80** Timothy Fuller; **83** MM; **85** Ken M. Johns/PR; **86** Michael Thomas/Stock South/PQ; **87** Dominic Oldershaw; **88** (t)Culver Pictures, (b)Hulton Getty Library/LA; **89** Aurthur Tilley/FPG; **90** (t)G. Brad Lewis/LA, (c)George B. Diebold/TSM, (b)Yoav Levy/Phototake/PQ; **91** (l)Macduff Everton/CB, (r)NASA/Mark Marten/PR; **92** Michael Thomas/Stock South/PQ; **94** Novastock/PE; **94-95** Cary Wolinsky/SB/PQ; **95** MM; **96** Dick Thomas/VU; **97** John Evans; **98** (t)Bob Woodward/TSM, (cl)Ping Amranand/Pictor, (cr)SuperStock, (b)Runk/Schoenberger from Grant Heilman; **99** Mark Thayer; **102** (l)Dr. Dennis Kunkel/PhotoTake NYC, (r)David Toase/PD; **104** (l)Bill Aron/PE, (r)Paul Silverman/FP; **105** (l)Digital Stock, (r)Joseph Pamieri/Pictor; **107** Geoff Butler; **109** Richard Megna/FP; **112** David M. Dennis; **113** David Young-Wolff/PE; **114 115** Roger Ressmeyer/CB; **117** Charles O'Rear/CB; **118** (t)MM, (b)Geoff Butler; **119** Geoff Butler; **120-121** Ed Welche's Antiques, Winslow, ME; **121** (t)courtesy Cheryl Landry, (b)The Stapleton Collection/Bridgeman Art Library; **122** (t)file photo, (c)MB, (b)Jeremy Horner/Stone; **124** Carol Christensen/Stock South/PQ; **128-129** PD; **130** (t)CB, (b)Artville; **131** (tl tr b)PD, (c)Artville, **132** (t c)Artville, (b)StudioOhio; **133** (t)PD, (c)Wolfgang Kaehler/CB, (b)CB; **134** Timothy Fuller; **138** Roger Ball/TSM; **140** (l)Geoff Butler, (r)Coco McCoy/Rainbow/PQ; **141** Dominic Oldershaw; **142** StudioOhio; **143** First Image; **145** MM; **148** Paul Barton/TSM; **151** Davis Barber/PE.

PERIODIC TABLE OF THE ELEMENTS

Columns of elements are called groups. Elements in the same group have similar chemical properties.

Element — Hydrogen
Atomic number — 1
Symbol — **H**
Atomic mass — 1.008
State of matter

Each element has a block in the periodic table. Within a block, you can find important information about the element.

1

1								
Hydrogen 1 **H** 1.008	**2**							
Lithium 3 **Li** 6.941	Beryllium 4 **Be** 9.012							
Sodium 11 **Na** 22.990	Magnesium 12 **Mg** 24.305	**3**	**4**	**5**	**6**	**7**	**8**	**9**
Potassium 19 **K** 39.098	Calcium 20 **Ca** 40.078	Scandium 21 **Sc** 44.956	Titanium 22 **Ti** 47.88	Vanadium 23 **V** 50.942	Chromium 24 **Cr** 51.996	Manganese 25 **Mn** 54.938	Iron 26 **Fe** 55.847	Cobalt 27 **Co** 58.933
Rubidium 37 **Rb** 85.468	Strontium 38 **Sr** 87.62	Yttrium 39 **Y** 88.906	Zirconium 40 **Zr** 91.224	Niobium 41 **Nb** 92.906	Molybdenum 42 **Mo** 95.94	Technetium 43 **Tc** 97.907	Ruthenium 44 **Ru** 101.07	Rhodium 45 **Rh** 102.906
Cesium 55 **Cs** 132.905	Barium 56 **Ba** 137.327	Lanthanum 57 **La** 138.906	Hafnium 72 **Hf** 178.49	Tantalum 73 **Ta** 180.948	Tungsten 74 **W** 183.84	Rhenium 75 **Re** 186.207	Osmium 76 **Os** 190.2	Iridium 77 **Ir** 192.22
Francium 87 **Fr** 223.020	Radium 88 **Ra** 226.025	Actinium 89 **Ac** 227.028	Rutherfordium 104 **Rf** (261)	Dubnium 105 **Db** (262)	Seaborgium 106 **Sg** (263)	Bohrium 107 **Bh** (262)	Hassium 108 **Hs** (265)	Meitnerium 109 **Mt** (266)

Rows of elements are called periods. Atomic number increases across a period.

The arrow shows where these elements would fit into the periodic table. They are moved to the bottom of the page to save space.

Lanthanide series

Cerium 58 **Ce** 140.115	Praseodymium 59 **Pr** 140.908	Neodymium 60 **Nd** 144.24	Promethium 61 **Pm** 144.913	Samarium 62 **Sm** 150.36

Actinide series

Thorium 90 **Th** 232.038	Protactinium 91 **Pa** 231.036	Uranium 92 **U** 238.029	Neptunium 93 **Np** 237.048	Plutonium 94 **Pu** 244.064